Communicating Chemistry

Patrick D. Bailey and Sara E. Shinton

Marjorie Cutter Scholarship

Royal Society of Chemistry

ROYAL SOCIETY OF CHEMISTRY

Communicating Chemistry

Compiled and developed by Patrick D. Bailey and Sara Shinton, Heriot Watt University
Designed by Imogen Bertin and Sara Roberts
Published by The Royal Society of Chemistry
Printed by The Royal Society of Chemistry

Copyright © The Royal Society of Chemistry 1999

For further information on other educational activities undertaken by the Royal Society of Chemistry write to:

The Education Department
Royal Society of Chemistry
Burlington House
Piccadilly
London W1V 0BN

Email: education@rsc.org

ISBN 0–85404–904–5

British Library Cataloguing in Data.

A catalogue for this book is available from the British Library.

Front cover background photograph by Sara Roberts
Front cover top small image – part of an illustration of the RSC Library and Information Centre at Burlington House
by Claire Milner-Johnson

Contents

Introduction

Communication skills are an essential part of all university degree courses, and chemistry is no exception. Recommendations from the National Committee of Inquiry into Higher Education, statements from industry, and feedback from recent chemistry graduates all emphasise the importance of communication skills. In general, communication skills are developed throughout undergraduate courses via a range of activities – for example, in tutorials and workshops, and during final year research projects – and many departments already have substantial parts of their course aimed at transferable skills. However, the exercises in this book can be used to address specific aspects of communication skills, or can be run as a complete 'module'.

There are two key themes underpinning the design of the book. Firstly, as communication skills are learnt rather than taught, the exercises provide students with many opportunities for first hand practice and experience. Secondly, the exercises are all set in a chemistry context, so students see the skills as interesting and relevant, and are encouraged to discover, explain and use chemistry. The aspects of communication skills identified in this pack are:

- information retrieval;
- written delivery;
- visual delivery;
- oral delivery;
- team work; and
- problem solving.

The material for each exercise is divided into three parts; a summary of the activity is given, including background information for tutors and a proposed timetable for running the exercise. This is followed by copies of the information for students, and then a detailed tutor's guide containing information needed for the tutor to run the exercise. In addition, suggestions of alternative ways/styles of running the exercise, sources of extra material and assessment ideas are included. This format ensures that the exercises can be run as described, or easily modified to meet specific needs. The activities are aimed at undergraduates in the penultimate year of a BSc(Hons) degree in chemistry. Brief descriptions of the ten exercises and their key features are given below, and detailed summaries of the exercises can be found at the start of each exercise.

Title of exercise	Key feature(s)
1 The Flurofen Problem	Team problem solving
2 Scientific Paper Workshop	Comprehension/problem solving
3 Computer Keyboard Skills	Basic computer skills
4 World Wide Web Treasure Hunt	Information retrieval
5 New Chemist Article	Writing a concise report
6 Dictionary of Interesting Chemistry	Information retrieval/ concise report writing
7 Hwuche-Hwuche Bark	Team work/problem solving
8 Annual Review Presentation	Oral presentations
9 Interviews and Interviewing	Interview skills
10 Poster Presentation	Preparing posters

The exercises typically require approximately two hours of contact teaching and ten hours total work from the students, although there is some variation. It is possible to run a "module" using several (or all) of the exercises, although each of the exercises presented in this book is stand-alone and concentrates on a particular aspect of communication skills.

The exercises have been trialled by colleagues at several institutions.

Patrick Bailey and Sara Shinton

Acknowledgements

The exercises in this pack were developed in 1996/97 with support from the Society's Marjorie Cutter Bequest. Much of the material is completely new, but some of it was taken from existing exercises run at Heriot-Watt and York Universities. All of the exercises described herein have been trialled with classes of more than forty students, and many friends have provided valuable feedback. These colleagues are acknowledged below – we apologise if we have inadvertently omitted anyone. We would welcome additional feedback from anybody who uses the material.

Dr John Garratt, who helped formulate many of the ideas in this book.

Colleagues at the University of York, who helped run some of the exercises in their early forms.

Chemistry colleagues at Heriot-Watt University, who have supported the introduction of the module on 'Communicating Chemistry', and for their help in running it.

Additional assistance from specific colleagues at Heriot-Watt University who have helped develop the material include Dr Kevin McCullough (computer skills), Dr Keith Morgan (Hwuche-Hwuche Bark), Dr Bill Steedman (data retrieval), Ms Catherine Fleming and Mr Malcolm Moffatt (library, data retrieval), Mr Nick Thow (careers service) and colleagues from the Learning & Teaching Centre. Dr Alan Boyd, Dr Rod Ferguson and Ms Christina Graham ran the spectra for Exercise 7.

Feedback has been much appreciated from the following people in other institutions:

Dr Jo-Ann Anderson (University of St Andrews)
Professor Mike Blandamer (University of Leicester)
Dr Bill Bentley (University of Wales, Swansea)
Dr Bernard Blessington (University of Bradford)
Dr Stephen Breuer (Lancaster University)
Dr Greig Chisholm (Ciba)
Dr Chris Cloke (University of Bath)
Dr Mike Cole (The Manchester Metropolitan University)
Dr Stephen Doughty (Oxford University)
Dr John Haddow (Bell College of Technology)
Dr Konstantinos Kalogerakis (Oxford University)
Dr Wolter Kaper (Faculteit Scheikunde, Amsterdam)
Dr Mary Masson (The University of Aberdeen)
Dr Bob Murray (Nottingham Trent University)
Dr Tina Overton (The University of Hull)
Dr Mike Turner (The University of Sheffield)
Dr Mark Winter (The University of Sheffield)
Dr Paul Wyeth (University of Southampton)
Dr Richard Young (The University of Newcastle upon Tyne)

We are particularly grateful for the support of the Royal Society of Chemistry – especially to Neville Reed for encouraging the initial idea, Denise Rafferty for overseeing the project, and for funding from the Marjorie Cutter Bequest.

Running the exercises

Each exercise includes an outline; a student handout denoted by "S"; and a tutor's guide denoted by "T". Additional points to note when running the exercises in this pack:

■ The tutor's guide and student handouts are designed to be sufficiently comprehensive to allow the exercises to be run with very little extra preparation. However, it is also possible to adapt them to meet specific course requirements. The exercises are aimed at undergraduates in their penultimate year of a BSc(Hons) course, but the material can be readily modified for use with first year students or postgraduates.

■ Each exercise can be run by a single tutor working with a class of students, although the involvement of two or more tutors for some exercises (especially in classes of more than forty) can be beneficial.

■ Some exercises are designed for students to work independently, while other exercises require students to work in groups or pairs. Group work is best carried out in groups of four to six, with work assigned to specific groups by the tutor before the start of the exercise.

■ The guidance notes assume that one hour teaching slots are available, and the suggested timetabling of exercises are based on fifty minute sessions of teaching/learning time.

■ Assessment of the exercises is not too difficult and practicable ways of doing so are included in the tutor's guide for each exercise.

■ Experience indicates that the exercises work best if they are a compulsory part of the course (ideally with course credit/assessment linked to them). However, students almost invariably undertake the exercises with great enthusiasm and commitment, achieving work of a high standard.

■ Each exercise can be run in isolation, to address a specific aspect of communication, or they can be run as a set of exercises to give a broader coverage of communication skills in chemistry.

The ten exercises can be run as a complete 'module', and there are some advantages in doing this. In particular, the importance placed on communication is emphasised, time can be saved by linking exercises together, and students can reinforce the skills they learn using several exercises. It can also be difficult to allocate enough time for developing transferable skills unless a specific module is identified for this purpose. The table on the following page shows the extent to which each aspect of communication skills is covered in the complete module based on this teaching pack. Feedback and self-assessment forms relating to the whole module are provided in Appendix E.

Summary of communication skills in each exercise

Title	Time* (Hours)	Information retrieval %	Written retrieval %	Visual delivery %	Oral delivery %	Team work %	Problem solving %
The Fluorofen Problem	1 (1)	0	0	0	0	50	50
Scientific Paper	3 (2)	20	0	0	0	30	50
Computer Keyboard Skills	10 (1)	0	10	90	0	0	0
World Wide Web Treasure Hunt	8 (1)	90	5	0	0	0	5
New Chemist Article	18 (1)	20	40	30	0	10	0
Dictionary of Interesting Chemistry	20 (1)	35	35	20	0	10	0
Hwuche-Hwuche Bark	8 (2)	5	0	10	15	45	25
Annual Review Presentation	12 (2.5)	10	0	20	70	0	0
Poster Presentation	12 (2)	20	25	40	0	15	0
Interviews	8 (2.5)	0	25	0	55	20	0
Overall percentage of module		20	14	21	14	18	13

* total time required by student; (tutor contact hours are given in brackets).

The Exercises

1. The Fluorofen Problem

Summary

Outline of the exercise

In this activity students are presented with a specific chemical problem set in an industrial context. The problem is based on improving the efficiency of a key step in the synthesis of Fluorofen, which is a pharmaceutical product whose patent will expire shortly. The initial handout sets the problem in context – each group of students represents a team of R&D chemists who work for a large company called ACE. Their competitors, Zenaxo, are intending to market Fluorofen at a reduced price when its patent runs out. The teams look at the synthesis of Fluorofen and decide where and how changes could be made to reduce their company's production costs. Some questions are outlined to direct them. The tutor's guide to this problem gives a step-by-step guide to running the hour-long workshop, which brings together aspects of practical organic chemistry, spectroscopic interpretation, mechanism and reaction kinetics.

Key aims

■ to introduce team working skills;

■ to introduce problem solving skills;

■ to develop awareness of industrial issues; and

■ to increase students' confidence in their ability to tackle realistic problems.

Time requirements

■ 1 hour workshop

■ No private study

Timetable

A proposed timetable for running this exercise within a one hour workshop is given below. The exercise is based around two main group discussion sessions, followed by plenary sessions in which ideas from all the groups are pooled.

Introduction	5 mins
Group discussions 1	15 mins
Plenary session 1	10 mins
Group discussions 2	15 mins
Plenary session 2	5 mins
Total	50 mins

S1

The Fluorofen Problem

Fluorofen student handout 1

What is Fluorofen?

■ Oral drug ■ Anti-inflammatory

■ Analgesic ■ Treatment for period pains

The ACE company makes Fluorofen

■ Your tutor is Head of Medicinal Chemistry at ACE

■ You are a crack team from R&D

The Problem

■ Patent expires in 6 months ■ Zenaxo will compete

■ They plan to undercut our post-patent price by 30% (£1.40 vs £2.00 for 100 tablets)

QUESTION 1: How might Zenaxo get their price so low?

Synthesis of Fluorofen

Key synthetic step

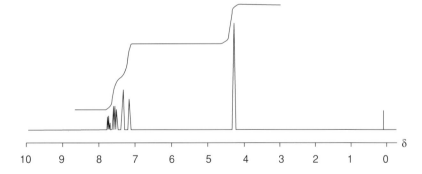

$M^+ = 159$

A
Expensive!

1. ⬡=O
2. HCl (aq)

D

Only 60% yield of D
+ by-product X (30% by weight)
Need double recrystallisation
to obtain pure D
uncontaminated by X

Mechanism for the formation of D

A

B

Mg

D

H⁺

C

Data on X
MS gives parent ion M^+
at 318

^{1}H NMR at 60 MHz:

δ

10   9   8   7   6   5   4   3   2   1   0

QUESTION 2: What is the structure of X?

QUESTION 3: Add arrows to the diagram of the formation of D to indicate how X is formed

QUESTION 4: What change in concentration of reactants would generate less X?

Higher [A]?

Higher [A] and [Mg]?

Higher [Mg]?

Higher [ketone]?

QUESTION 5: How would you achieve your aim in Q4?

Hint: There are many ways Zenaxo could undercut ACE. Your team would probably want to:

■ Optimise reaction conditions

■ Economise on solvents, reactants, running costs (time, heat) and waste disposal

You may also want to look at other work on Grignard reactions such as that by John Brown's group in Oxford – they went back to the literature to find clues[1] to improving "difficult Grignard reactions" and carried out a careful study to provide a general procedure[2]. Do you think that this solution would work well for the Fluorofen problem?

[1]A Mendel, *J. Organomet. Chem.*, 1966,**6**, 97.
[2]K.V. Baker, J.M. Brown, N. Hughes, A.J. Skarnulis and A. Sexton, *J. Org. Chem.*, 1991, **56**, 698.

The Fluorofen problem

This exercise is particularly effective if run entirely 'in character', with the tutor taking on the role of Head of Medicinal Chemistry throughout the workshop. It may be useful to follow the detailed timetable for each part of the workshop (see Summary), so that the pace can be pushed if necessary – a sense of urgency helps the exercise to be successful and is particularly important if the timetable does not permit the workshop to overrun.

No preparation is required by students. This has two advantages; no assumptions need to be made that students have carried out background reading, and the exercise has immediate interest and impact. The following guidance notes relate to the timetable in the summary at the start of this exercise.

Introduction

Addressing the whole class in a role-play is an effective way of introducing the activity. Possible points to include in an introductory presentation are given below and can be reinforced by using Handout 1. Alternatively, the students can be divided immediately into groups, and simply provided with the information on the handout.

■ **Introduction**
In the role play scenario the tutor acts as head of medicinal chemistry of the company ACE, and the students as a top R&D team.

■ **Why has the meeting been called?**
This urgent and important meeting has been called because the patent of one of the company's leading pharmaceutical products, Fluorofen, will expire in six months and the company is worried about competition from other manufacturers. They have already been informed that Zenaxo is planning to undercut their price by 30%.

■ **What does the company plan to do?**
The customer base cannot be lost, and therefore ACE plans to match Zenaxo's price for a few months at zero profit, while working to cut costs.

■ **Information about Fluorofen (Handout 1).**
Fluorfen has anti-inflammatory and analgesic properties. It is excellent for treating period pains, and has a large and sustainable market worth about £50 million per annum in sales. The structure of Fluorofen and a synthetic sequence for its manufacture are shown on Handout 1.

■ **What do the teams have to do?**
The company feels that a fresh look at the problem is needed and has set up some meetings to brainstorm the problem. The groups therefore have fifteen minutes in which to come up with five or six reasons why Zenaxo may be able to undercut their price.

Group discussions 1

Assigning students to defined groups, rather than letting them choose their own groups, works well. One option is to assign them to 'companies' that are subsidiaries of ACE, and to ask them to invent a company name. Once they have organised themselves into groups, and have received Handout 1, it is important that they identify a spokesperson, and start to write down ideas – each group must be asked for suggestions when the fifteen minutes are up. If a group is stuck or slow, the suggestions below can be used to guide them.

Plenary session 1

Suggestions for how Zenaxo might be undercutting the companies can be pooled by picking on various groups. Some examples are given in the table below.

Some suggestions for cutting the cost of Fluorofen production

General ideas	Specific ideas
Lower non-chemistry production costs	Cheaper starting materials Less on packaging/marketing
Lower production costs (chemistry) · · · · · ·	Better reaction conditions: – Temperature – Pressure – Solvent – Reaction time – Cheaper reagents (*eg* catalyst) More efficient purification Recycling: – Reactants – Solvents – By-products Improved yield(s)
Cheaper route	Completely new route Alternative key step Cheaper analogue of Fluorofen
Someone is cheating!	Zenaxo selling Fluorofen at a loss while they corner the market.

A brief discussion session at the end of this part can be used to eliminate suggestions that are not practicable. At the end of the plenary session, the results of the brainstorming session should be summarised including the following points (either verbally, or on an additional handout):

■ The most likely source of cost cuts is through improvement of the efficiency of the Grignard reaction. In particular, the students might have identified the following:
– High cost of the starting material
  (but note that it probably cannot be bought or made more cheaply)
– Poor yield
– High running costs (long reaction time and high purification costs)

■ If there is time, students can be asked for suggestions about how the reactions might be made to take place more quickly and cleanly in order to reduce costs.

■ It could be suggested to students that, rather than changing conditions by trial and error, they might try to identify impurities in an attempt to design conditions that would reduce the formation of the by-product(s) (see Handout 2).

At the end of the first plenary session, Handouts 2 and 3 should be distributed and students must work through the questions.

Group discussions 2

Groups should need little tutor input now. The by-product is the dimer of CF_3–C_6H_4–CH_2, and it is quite easy to come up with a plausible mechanism (see below). (Although a multi-step electron transfer mechanism is very likely the details of the mechanism do not affect the subsequent kinetic argument or solution to the problem, and therefore in depth discussion of it depends on the amount of time available). The tutor need only guide those groups that are getting behind. Group discussions work well if one company is asked to put forward the structure of X (once everyone has determined it), and another is asked to suggest the mechanism for its formation (perhaps adding arrows to Handout 2). At least one group should have an idea for the answer for question 5 (see Handout 3) before the final plenary session.

The structure of X, and a possible mechanism for its formation.

Plenary session 2

This works best as an interactive session. However, a possible summary, which could be presented in the form of a memo from the company, is given below.

MEMO

From: Head of Medicinal Chemistry To: R&D Team
Subject: Flurofen synthesis costs

Cutting the cost of Flurofen
The main problem is the low yield for step 1, due to the formation of a by-product X.

How can we compete?
The structure of **X** and a mechanism for its formation have been identified. If the [Mg] is increased, the formation of the Grignard would be quicker and there would be less time for the dimer by-product to form before all of the starting material is consumed. This would also cut running costs, as the reaction time would drop, and the purification of the product should be easier.

How could the [Mg] be increased?
Suggestions include using a powdered form of magnesium, using a thin film or precipitating/depositing it onto a porous material. However, the 'dry-stir' method (reference 2, Handout 3) will probably solve the Fluorofen problem. This is a cheap and efficient way of introducing magnesium with a high surface area. Using high surface area magnesium and general optimisation of the reaction should allow us to reduce the cost by around 50%.

Adapting/extending the exercise

The scenario of a company needing to improve the efficiency of a synthesis provides a useful backdrop for a number of different chemical problems that could be matched to course content and undergraduate level. *The Fluorofen Problem* requires a range of chemical skills at modest level, but the exercise could be biased towards a particular area of chemistry, if that were deemed more appropriate – *eg* practical problems (including industrial factors relating to scale-up), structure solving from spectra, mechanisms, physical organic chemistry, or literature searching. If several of these aspects were followed in more detail, further workshop time would be required, or the students would need to carry out some private study; the latter option offers an easy way of carrying out an assessment.

Possible extensions:

■ Ask the companies for a brief report on possible ways of saving money on the synthesis;

■ Provide data on by-products from several of the steps so that the exercise has a larger component of structure determination (which could be assessed);

■ Ask for a literature search, in order to:
  – find ways of producing high surface area magnesium
  – find the specific references relating to the exercise (see Handout 3); or

■ Link in a subsequent laboratory experiment, in which (on a simpler/cheaper benzyl derivative) students compare yields from old and new synthetic routes.

Assessment

This exercise works particularly well as an ice-breaker for subsequent team work, and there may be no need to generate a specific mark from this workshop. However, it can form part of peer group assessment of team working skills (see Appendix E).

Other methods of assessing the exercise are to ask for written work to be handed in as part of an extension to the exercise (either from individuals, or from the teams). For example through:

■ the production of a report summarising the ways in which the synthesis of Fluorofen might be made more cost effective. Reports could be approximately 300 words, and might include a synthetic scheme. As many sources of savings as possible should be identified.

■ the production of a report of approximately 300 words summarising how high surface area magnesium might be formed. Reports could briefly explain why this is of relevance to *The Fluorofen Problem,* and identify six literature references, including a 1966 paper by Mendel *et al.,* and a 1991 paper by Baker *et al.* (see Handout 3).

2. Scientific paper workshop

Summary

Outline of the exercise

In this exercise students analyse and discuss in groups a short, recently published research paper. They are presented with the paper in short sections on individual handouts; a number of questions follow each section, and are designed to check students' understanding and interpretation of what they have read and to elicit ideas about how to tackle specific practical problems. The exercise is an ideal introduction to handling scientific literature, and it aims to build students' confidence in reading and interpreting papers. The emphasis on effective team work throughout the exercise should also be noted.

This particular article is based on the following paper: "Direct Proof of the Involvement of a Spiro Intermediate in the Pictet-Spengler reaction", P.D. Bailey, *J. Chem. Res.*, 1987, 202–203. General guidance for the preparation of similar exercises based on other papers are provided in the tutor's guide.

Key aims

- to foster team working skills;

- to improve problem solving skills;

- to introduce information retrieval skills;

- to increase students' confidence in their ability to comprehend primary literature; and

- to plan and interpret (modest) research experiments.

Time requirements

- 2 hour workshop (tutor contact time)

- 1 hour private study

A proposed timetable for the exercise is given below, and is based on two 1 hour teaching slots. Groups need to be given the handout for section 3 at the end of the first workshop, so that they can spend approximately an hour studying and discussing it before Workshop 2.

Workshop 1		Workshop 2	
Introduction	10 mins	Section 3 (groups)	10 mins
Section 1 (groups)	10 mins	Section 3 (plenary)	5 mins
Section 1 (plenary)	5 mins	Section 4 (groups)	5 mins
Section 2 (groups)	10 mins	Section 4 (plenary)	5 mins
Section 2 (plenary)	15 mins*	Section 5 (groups)	10 mins
		Section 5 (plenary)	5 mins
(*Hand out Section 3 after workshop 1)		Section 6 (groups)	5 mins
		Section 6 (plenary)	5 mins
Total	50 mins		50 mins

Scientific paper workshop

Student handout 1

Investigating the Pathway taken by the Pictet-Spengler Reaction

This exercise is based on the following paper: P. D. Bailey, *J. Chem. Res.* (S), 1987, 202–203

Background information

Many naturally occurring alkaloids contain the tetrahydro-β-carboline unit:

Such indolic natural products, and derivatives of them, can be synthesised in the laboratory by the Pictet-Spengler reaction, which involves condensation between an amine and an aldehyde, followed by ring closure. The reaction scheme below shows one example of a Pictet-Spengler reaction involving a familiar indolic reactant. The naturally occurring amino acid L-tryptophan, **A**, reacts to form a 1,3-disubstituted tetrahydro-β-carboline **B**; notice that **B** is formed as a mixture of *cis* and *trans* isomers.

This workshop is based on a short paper published in a chemical journal. The work described in the paper addressed an important and challenging chemical problem, but the answer to the problem required simple chemical procedures with which you are familiar. By the end of the exercise you will have had an opportunity to read, discuss and comment on the whole paper, and the work that it describes.

Introduction

■ Read the introduction to the paper. (The small numbers shown in superscript refer to published papers which justify the statements made by the authors.)

Direct Proof of the Involvement of a Spiro Intermediate in the Pictet-Spengler Reaction

Recent syntheses of several alkaloids have relied upon the stereospecific formation of tetrahydro-β-carbolines via the Pictet-Spengler reaction[1]: in particular, both *cis*- and *trans*-1,3-disubstituted derivatives have been used in asymmetric routes to a number of indolic natural products.[1a–c] If the stereochemical control possible using the Pictet-Spengler reaction is to be fully exploited, it is vital that the reaction pathway should be elucidated; we therefore undertook to carry out a detailed study on the mechanism of this reaction.

■ Now answer the questions below, which help you think about why the author was interested in this particular problem.

1. What are alkaloids?

2. Why should anyone want to synthesise alkaloids?

3. What features of a synthesis would make it a 'good' synthesis?

4. Why might an understanding of the mechanism of the Pictet-Spengler reaction help to achieve a 'good' synthesis of an indolic alkaloid?

Student Handout 2
Methods for following reaction pathways

■ Read the second paragraph of the paper, which gives information about the state of knowledge of possible reaction pathways at the time this work was started.

There are two main pathways (Scheme 1) by which the ring closure could take place, involving either direct attack at the indole 2-position (route a), or attack at the 3-position followed by migration (route b).[2]

Experiments on related systems have suggested that a spiroindolenine intermediate (**4**) is probably involved (route b);[3,4] however, electrophilic attack at the indole 2-position is known to compete (in acyclic systems) with attack at the 3-position,[5] and it has been noted that certain stereochemical features of the Pictet-Spengler reaction are consistent with either mechanism;[6] moreover, attack at the indole 3-position would presumably involve the 'disfavoured' 5-*endo-trig* ring-closure, whereas direct attack at the 2-position would proceed through the 'favoured' 6-*endo-trig* pathway.[7]

1. Does the author give you the impression that route a is more likely than route b, or vice versa, or that either route is equally likely, or that both routes occur simultaneously? From the way the author has presented the case, suggest likely odds from a bookmaker for the Pictet-Spengler reaction proceeding via route a or route b.

2. Many different methods have been used to study reaction pathways – list as many as you can (general methods and specific techniques). Decide whether these techniques could be used to investigate which of the two suggested routes for the Pictet-Spengler reaction is correct.

Student Handout 3
Examining the author's strategy

■ Read the following summary of the author's plans.

In an attempt to clarify the mechanism, we studied the formation of 2,3-dimethyl-1,2,3,4-tetrahydro-3-aza-β-carboline (**7**) by the condensation of the indolic hydrazine (**6**) with formaldehyde, so that any spiro intermediate would possess a plane of symmetry. On repeating the experiment using isotopically labelled formaldehyde, we expected that the label would have been localised on C(1) if direct attack had occurred at the indole 2-position (route a), but would have been distributed between C(1) and C(4) if migration from the indole 3-position had occurred (route b).

6	**7**
reactant	product

1. What general technique did the author actually use?

2. Draw structures of the intermediates you would expect to obtain in the formation of 2,3-dimethyl-1,2,3,4-tetrahydro-3-aza-β-carboline (**7**) if the reaction proceeds via:

   (i) Route a (*ie* the 3-aza-analogue of **3**);

   (ii) Route b (*ie* the 3-aza-analogue of **4**).

   There is something special about the intermediate from (ii), which might allow the pathways to be distinguished – what is it? (Note: you might find this much easier to see if you make a simple molecular model of the intermediate).

3. Hence, explain why the author expected to be able to clarify the mechanism by carrying out the reaction using labelled methanal, and then studying the distribution of the isotopic label in the final product.

4. Formaldehyde could be labelled with ^{2}H, ^{3}H, ^{13}C, ^{14}C, ^{17}O, or ^{18}O. No labelled formaldehyde was commercially available at the time this experiment was done, so the author had to make some. Consider each of the isotopes and decide how you would determine their positions in the final product. Which isotope do you think would most easily provide an answer?

Student Handout 4
The synthesis of products

■ Read what the author did next.

The indolic hydrazine (**6**) was prepared by the reaction of gramine methosulphate[8] with 1,2-dimethylhydrazine in the presence of aqueous NaOH;[9] reaction of (**6**) with aqueous formaldehyde in methanolic HCl gave the expected Pictet-Spengler adduct (**7**) (26%) and its N^{in}-methoxymethyl derivative (**8**) (50%), after purification by flash chromatography.[10] [^{2}H$_2$]Formaldehyde was generated by oxidising methanol with pyridinium chlorochromate (PCC),[11] and the CD$_2$O was swept by nitrogen into a solution of the hydrazine (**6**) in CD$_3$OD–HCl–D$_2$O. The condensation reaction proceeded as expected, the tetrahydro-3-aza-β-carboline being isolated as before: its isotopic composition was investigated.

1. Which isotope did the author opt for? Why do you think this isotope was chosen?

2. Why do you think the author initially carried out the synthesis in the absence of an isotope?

3. What is the origin of compound **8** which was found in 50% yield? (In other words, what are the sources of the tricycle (**X**), the CH$_2$ (**Y**) and the MeO (**Z**) in structure **8**?)

4. The author is apparently satisfied with a yield of the desired product (**7**) of only 26%. Why?

5. Decide which technique(s) you would use to investigate the isotopic composition of compound **7** and give reasons for your choice.

Student Handout 5
Analysis of results

■ Here is how the authors studied the isotopically labelled product 7:

The condensation reaction proceeded as expected, the tetrahydro-3-aza-β-carboline being isolated as before: its isotopic composition was investigated using ¹H NMR spectroscopy and mass spectrometry.

■ Read the extracts that describe the results, and answer the questions below which help you to interpret them.

The NMR spectrum

The N-methyl groups, present as two sharp singlets in the ¹H NMR spectrum of the non-deuterated compound, were each resolved into four singlets of roughly equal intensity in the deuterated product. (Figure 1).

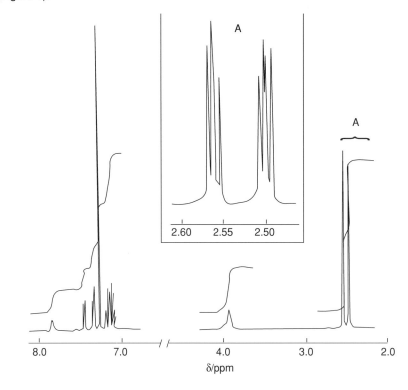

Figure 1 360 MHz ¹H NMR spectrum of the deuterated Pictet-Spengler adduct

The C(1) and C(4) hydrogens would be expected to resonate at approximately δ4 in unlabelled **7**.

1. Explain how you might expect to be able to use ¹H NMR to tell whether the product had deuterium atoms at positions C(1) and/or C(4)? (In other words what *specific* change in the NMR of the labelled product *vs* the unlabelled product should allow the pathway to be elucidated?)

2. The C(1) and C(4) protons are broad and poorly resolved in the ¹H NMR spectrum of the undeuterated adduct. Why can these signals not be used to determine the location of deuterium in the deuterated adduct?

3. The N-methyl hydrogens resonate in the region around δ2.5. Explain why the two methyl groups give signals at slightly different chemical shifts, and why both signals were present as sharp singlets in the undeuterated adduct.

4. Do you agree that there are four signals in the ^{1}H NMR for each of the N-methyl groups?

5. The deuterated compound gave rise to a spectrum containing 'four singlets of roughly equal intensity in the region of δ2.5'. Look at the structure below representing the deuterated product, and decide what arrangement of hydrogen and deuterium in the positions X and Y could give rise to this observation. (Note that the CH_2/CD_2 groups remain intact; isotopes cause slight changes in chemical shifts).

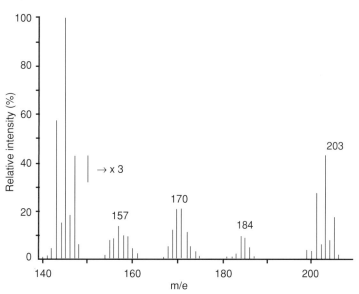

The mass spectrum

Moreover, the mass spectrum revealed parent ions not only at the expected m/z of 203, but also at 201 and 205 (Figure 2).

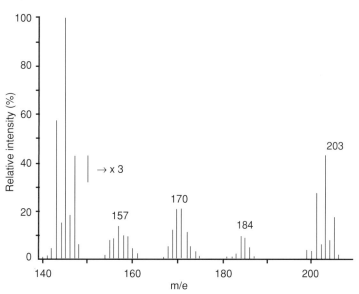

Figure 2 Mass spectrum (E.I.) of the deuterated Pictet-Spengler adduct.
Relative ratio of peaks at 201:203:205 = 30:48:22

6. Suggest the identities of the molecules which give rise to the parent ions which appear at m/z 201, 203 and 205 in the spectrum.

7. Does your interpretation of the mass spectrum support your ideas about the identity of the deuterated product that you formulated from the NMR spectrum?

Student handout 6
The author's conclusions

The author concluded that the 4 NMR signals for each methyl group at δ2.5 were due, not to coupling, but to the presence of four separate compounds which were very similar indeed:

It was therefore concluded that the labelled product consisted of roughly equal proportions of (**9**), (**10**), (**11**) and (**12**).

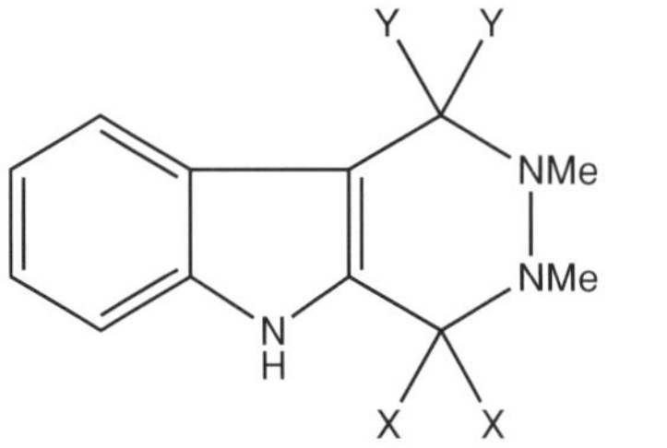

(9) X=Y=H
(10) X=D, Y=H
(11) X=H, Y=D
(12) X=Y=D

1. Did you reach the same conclusion as the author?(If not, who is right?)

2. Are these results consistent with pathway a, or pathway b, or neither?

■ Now read the author's conclusions about the pathway taken by the Pictet-Spengler reaction.

This statistical mixing of the deuterium label is consistent with the series of equilibria shown in Scheme 2, involving a spiro intermediate and reversible imine formation-hydrolysis; subsequent cyclisation to the tetrahydro-3-aza-β-carboline (**7**) must have been slow with respect to these processes.

Therefore, we have provided unambiguous evidence that this Pictet-Spengler reaction involves the rapid reversible formation of a spiro intermediate, and thus it seems likely that tetrahydro-β-carbolines are formed by a migratory pathway. We believe that the findings described herein could lead to a better understanding of the stereochemical features of the Pictet-Spengler reaction.

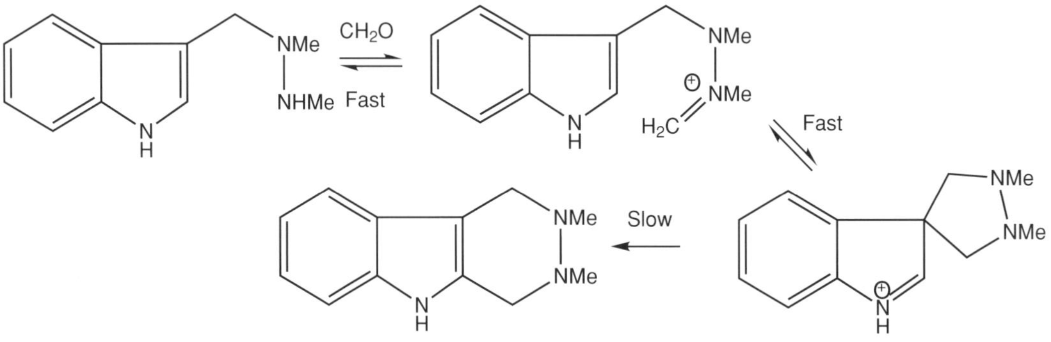

Scheme 2. Proposed mechanism for the Pictet-Spengler reaction

3. The most unexpected observation was the formation of a product containing four deuterium atoms – convince yourself that Scheme 2 would allow this to occur, and hence that all of the products **9–12** could have been formed if the mechanism in Scheme 2 were operating.

■ Summarise, in your own words, what the author claims to have discovered about the pathway taken by the Pictet-Spengler reaction.

Your summary of the conclusion from the work (maximum 50 words):

_____

_____

_____

_____

_____

_____

_____

The author wrote a fifty-word abstract as the first paragraph of the paper – have a look at that summary. Do you think it is clear and accurate?

Some interesting observations... and what happened later

a) There seems to be little doubt that the author would have used ^{13}C as the label of choice, but presumably selected deuterium on the grounds of cost. He almost certainly expected the loss of ^{1}H signals in the ^{1}H NMR to be used to identify the pathway, and was disappointed to find that it could not be used reliably. But he probably did not expect the isotope effect to be sufficient to yield the necessary data. The mass spectrometry was very important, in order to back up the interpretation. If the isotope effect had not distinguished the methyl groups in the ^{1}H NMR, he could have resorted to the use of a ^{13}C label, although there are other ways to determined the location of the deuterium atoms (how?).

b) The model reactions were carried out in aqueous solution, because (unlabelled) methanal comes as an aqueous solution (formalin; CH_2O is a gas that readily polymerises). With the conditions sorted out, the labelled reaction was also carried out under these conditions. Because imine formation is reversible _in the presence of water_, it became clear that formation of the spiro intermediate was reversible (see c). If water had not been present (and Pictet-Spengler reactions are usually conducted under anhydrous conditions), the involvement of the spiro intermediate would have been proven, but there would have been no evidence that its formation was reversible (**10** and **11** would be formed in equal amounts, whether it was reversible or not).

c) Because the formation of the spiro intermediate was fast and reversible, it was concluded that six-membered ring formation (_via_ route a or route b – we cannot tell which) must be the slow, rate determining step, and that factors controlling the stereochemistry of six-membered ring formation should operate. The research group was able to show a few years later that this conclusion was correct[1] – the stereochemistry of the Pictet-Spengler reaction can be controlled by careful choice of substituents and conditions, exactly as predicted by the pathway in Scheme 2. In particular, reference 1 provides a general method of making _cis_-1,3-disubstituted tetrahydro-β-carbolines, which has now been used to synthesise indole alkaloids such as suaveoline[2].

[1] P. D. Bailey _et al._, _J. Chem. Soc., Perkin Trans. 1_, 1993, 431–9.

[2] P. D. Bailey and K. M. Morgan, _J. Chem. Soc., Chem. Commun._, 1996, 1479–80.

Direct Proof of the Involvement of a Spiro Intermediate in the Pictet–Spengler Reaction†

J. Chem. Research (S),
1987, 202–203†

Patrick D. Bailey

Department of Chemistry, University of York, Heslington, York YO1 5DD, U.K.

The formation of 2,3-dimethyl-1,2,3,4-tetrahydro-3-aza-β-carboline (**7**)‡ by the Pictet–Spengler reaction is shown herein to involve the rapid reversible formation of a spiro-indolenine intermediate, giving direct proof of the involvement of such a species in this reaction.

Recent syntheses of several alkaloids have relied upon the stereospecific formation of tetrahydro-β-carbolines *via* the Pictet–Spengler reaction:[1] in particular, both *cis*- and *trans*-1,3-disubstituted derivatives have been used in asymmetric routes to a number of indolic natural products.[1a-c] If the stereochemical control possible using the Pictet–Spengler reaction is to be fully exploited, it is vital that the reaction pathway should be elucidated; we therefore undertook to carry out a detailed study on the mechanism of this reaction.

There are two main pathways (Scheme 1) by which the ring closure could take place, involving either direct attack at the indole 2-position (route a), or attack at the 3-position

the label would have been localised on C(1) if direct attack had occurred at the indole 2-position (route a), but would have been distributed between C(1) and C(4) if migration from the indole 3-position had occurred (route b).

The indolic hydrazine (**6**) was prepared by the reaction of gramine methosulphate[8] with 1,2-dimethylhydrazine in the presence of aqueous NaOH;[9] reaction of (**6**) with aqueous formaldehyde in methanolic HCl gave the expected Pictet–Spengler adduct (**7**) (26%) and its N^{in}-methoxymethyl derivative (**8**) (50%), after purification by flash chromatography.[10] [²H₂]Formaldehyde was generated by oxidising methanol with pyridinium chlorochromate (PCC),[11] and the CD₂O was swept by nitrogen into a solution of the hydrazine (**6**) in CD₃OD–HCl–D₂O. The condensation reaction proceeded as expected, the tetrahydro-3-aza-β-carboline being isolated as before: its isotopic composition was investigated by n.m.r. spectroscopy and mass spectrometry.

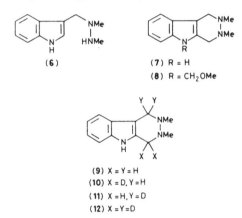

(**6**)

(**7**) R = H

(**8**) R = CH₂OMe

(**9**) X = Y = H
(**10**) X = D, Y = H
(**11**) X = H, Y = D
(**12**) X = Y = D

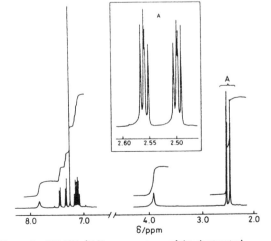

Scheme 1   Possible pathways for the Pictet–Spengler reaction

followed by migration (route b).[2] Experiments on related systems have suggested that a spiroindolenine intermediate (**4**) is probably involved (route b);[3,4] however, electrophilic attack at the indole 2-position is known to compete (in acyclic systems) with attack at the 3-position,[5] and it has been noted that certain stereochemical features of the Pictet–Spengler reaction are consistent with either mechanism;[6] moreover, attack at the indole 3-position would presumably involve 'disfavoured' 5-*endo-trig* ring-closure, whereas direct attack at the 2-position could proceed through the 'favoured' 6-*endo-trig* pathway.[7]

In an attempt to clarify the mechanism, we studied the formation of 2,3-dimethyl-1,2,3,4-tetrahydro-3-aza-β-carboline (**7**) by the condensation of the indolic hydrazine (**6**) with formaldehyde, so that any spiro intermediate would possess a plane of symmetry. On repeating the experiment using isotopically labelled formaldehyde, we expected that

It was immediately apparent that the spectra did not correspond to either of the expected results. The *N*-methyl groups, present as two sharp singlets in the ¹H n.m.r. spectrum of the non-deuterated compound, were each resolved into four singlets of roughly equal intensity in the deuterated product (Figure 1). Moreover, the mass spectrum revealed parent ions not only at the expected *m/z* of 203, but also at 201 and

---

†This is a Short Paper as defined in the Instructions for Authors [*J. Chem. Research (S)*, 1987, Issue 1, p. ii]; there is therefore no corresponding material in *J. Chem. Research (M)*.
‡To facilitate the discussion and comparison with previous work on β-carbolines, the term '3-aza-β-carboline' is used for a β-carboline in which the CH at position 3 has been replaced by N. The systematic name for compound (**7**) is given in the Experimental section.

Figure 1   360 MHz ¹H N.m.r. spectrum of the deuterated Pictet–Spengler adduct

205 (Figure 2). It was therefore concluded that the labelled product consisted of roughly equal proportions of (9), (10), (11), and (12). This statistical mixing of the deuterium label is consistent with the series of equilibria shown in Scheme 2, involving a spiro intermediate and reversible imine formation–hydrolysis; subsequent cyclisation to the tetrahydro-3-

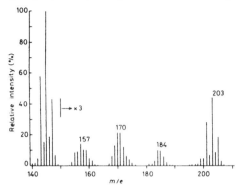

Figure 2 Mass spectrum (e.i.) of the deuterated Pictet–Spengler adduct. Relative ratio of peaks at 201 : 203 : 205 = 30 : 48 : 22

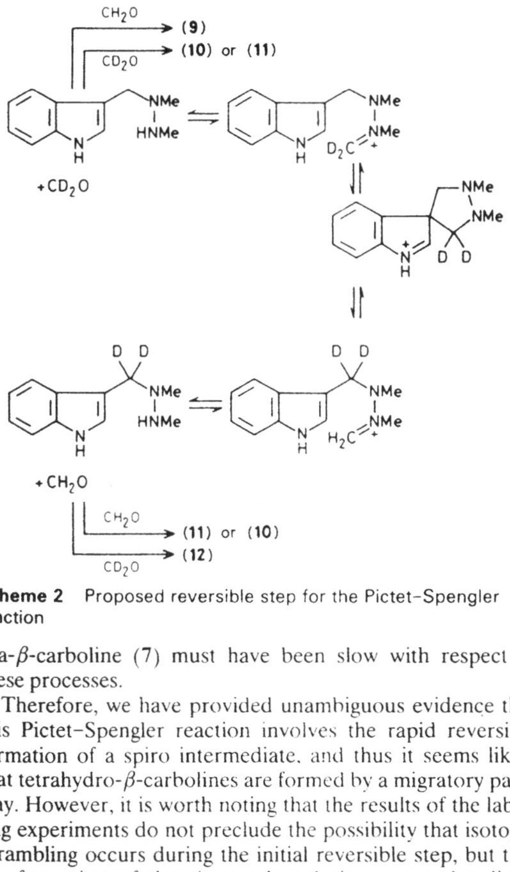

Scheme 2 Proposed reversible step for the Pictet–Spengler reaction

aza-β-carboline (7) must have been slow with respect to these processes.

Therefore, we have provided unambiguous evidence that this Pictet–Spengler reaction involves the rapid reversible formation of a spiro intermediate, and thus it seems likely that tetrahydro-β-carbolines are formed by a migratory pathway. However, it is worth noting that the results of the labelling experiments do not preclude the possibility that isotopic scrambling occurs during the initial reversible step, but that the formation of the six-membered ring occurs by direct attack at the indole 2-position. Nevertheless, we believe that the findings described herein could lead to a better understanding of the stereochemical features of the Pictet–Spengler reaction.

Experimental

M.p.s were determined on a Reichert microscope hot-stage apparatus. N.m.r. spectra were recorded on a JEOL FX90Q spectrometer at 90 MHz (^{1}H) or 22.5 MHz (^{13}C), and chemical shifts are quoted in ppm downfield from Me$_4$Si as internal standard. Mass spectra were obtained by electron impact at 70 eV on an A.E.I. MS3074 spectrometer. All solvents were purified and dried by standard methods. Flash chromatography[10] was carried out using silica as the stationary phase.

1-(*Indol-3-ylmethyl*)-1,2-*dimethylhydrazine* (**6**).—Gramine methosulphate[8] (12.0 g, 40.0 mmol) and 1,2-dimethylhydrazine dihydrochloride (5.32 g, 40.0 mmol) were dissolved in 30% aqueous MeOH (30 ml) at 0 °C and 2 M aqueous NaOH (80 ml) was added. After stirring for 0.5 h at 0 °C, and 1.75 h at room temperature, the mixture was extracted into CH$_2$Cl$_2$ (200 ml total). After removal of the solvent, the product was purified by flash chromatography (MeOH–CHCl$_3$; 1 : 4 v/v), giving a golden oil (4.98 g, 66%); δ_H (CDCl$_3$) 2.41 (3 H, s), 2.60 (3 H, s), 3.02 (1 H, br s), 3.96 (2 H, s), 6.91–7.74 (5 H, m), and 9.49 (1 H, br s); δ_C (CDCl$_3$) 34.83 (q), 43.29 (q), 52.28 (t), 109.43 (s), 111.33 (d), 119.08 (d), 119.35 (d), 121.68 (d), 124.76 (d), 127.96 (s), and 136.30 (s); m/z 189 (M$^+$), 130, and 60 (Found: M$^+$, 189.1265. C$_{11}$H$_{15}$N$_3$ requires M, 189.1266).

2,3-*Dimethyl*-2,3,4,5-*tetrahydro*-1H-*pyridazino*[4,5-b] *indole* (7).—The indolic hydrazine (**6**) (95 mg, 0.50 mmol) in a mixture of MeOH (10 ml) and 2 M HCl (0.25 ml) was reacted with 40% w/v aqueous CH$_2$O (0.1 ml, 1.2 mmol) at 50 °C under N$_2$ for 3 h. After removal of the methanol *in vacuo*, the product was taken up in CH$_2$Cl$_2$, washed with 1 M aqueous NaOH, and evaporated to dryness. Preparative t.l.c. (Chromatotron) in MeOH–CHCl$_3$ (1 : 9 v/v) eluted 5-*methoxymethyl*-2,3-*dimethyl*-2,3,4,5-*tetrahydro*-1H-*pyridazino*[4,5-b] *indole* (**8**) (62 mg, 50%) as the higher R_f component; δ_H (CDCl$_3$) 2.52 (3 H, s), 2.55 (3 H, s), 3.19 (3 H, s), 3.92 (2 H, br s), 3.98 (2 H, br s), 5.34 (2 H, s), and 7.09–7.49 (4 H, m); δ_C (CDCl$_3$) 35.70 (q), 37.87 (q), 46.48 (t), 47.73 (t), 55.64 (q), 73.95 (q), 106.99 (s), 109.38 (d), 117.99 (d), 120.00 (d), 121.68 (d), 126.33 (s), 130.99 (s), and 137.33 (s); m/z 245 (M$^+$), 187, 157, and 45. (Found: M$^+$, 245.1530. C$_{14}$H$_{19}$N$_3$O requires M, 245.1528). The desired product (7) was a component of lower R_f and was obtained as a golden oil (26 mg, 26%), which could be crystallized from CH$_2$Cl$_2$ giving colourless *crystals*, m.p. 163.5–164.5 °C; δ_H (CDCl$_3$) 2.40 (3 H, s), 2.53 (3 H, s), 3.71 (2 H, br s), 3.91 (2 H, br s), 7.03–7.47 (4 H, m), and 8.68 (1 H, br s); δ_C (CDCl$_3$) 35.33 (q), 37.83 (q), 46.60 (t), 48.44 (t), 105.49 (s), 111.12 (d), 117.62 (d), 119.30 (d), 121.31 (d), 125.91 (s), 129.49 (s), and 136.20 (s); m/z 201 (M$^+$), 143, and 43 (Found: M$^+$, 201.1271. C$_{12}$H$_{15}$N$_3$ requires M, 201.1266).

Deuterated Pictet–Spengler Reaction.—The CD$_2$O was generated by the reaction of PCC[11] (2 g) in CH$_2$Cl$_2$ with CD$_3$OD (2 ml) under reflux; the CD$_2$O was flushed into the reaction vessel by a continuous stream of N$_2$. The indolic hydrazine (**6**) (51 mg, 0.27 mmol) was dissolved in a mixture of D$_2$O (6 ml), 12 M HCl (0.25 ml), and CD$_3$OD (5 ml), and was then allowed to react with the CD$_2$O during 3 h at 80 °C under N$_2$. After removal of the solvents *in vacuo*, the deuterated Pictet–Spengler adducts (9)–(12) were isolated by flash chromatography (MeOH–CHCl$_3$; 1 : 9 v/v) as a colourless oil (4.8 mg, 8.8%). which was identical with the non-deuterated parent (7) by t.l.c., and which gave the ^{1}H n.m.r. and mass spectra shown in Figures 1 and 2 respectively (Found for M$^+$ at 205: 205.1514. C$_{12}$H$_{11}$D$_4$N$_3$ requires M, 205.1517. Found for M$^+$ at 203: 203.1394. C$_{12}$H$_{13}$D$_2$N$_3$ requires M, 203.1392).

Received, 17th November 1986; Paper E/263/86

References

1 (a) G. Massiot and T. Mulamba, *J. Chem. Soc., Chem. Commun.*, 1983, 1147; 1984, 715; (b) P. Flecker and E. Winterfeldt, *Tetrahedron*, 1984, **40**, 4853; (c) M. Shimizu, M. Ishikawa, Y. Komoda, T. Nakajima, K. Yamaguchi, and S. Sakai, *Chem. Pharm. Bull.*, 1984, **32**, 1313; (d) T. Suzuki, E. Sato, K. Unno, and T. Kametani, *Heterocycles*, 1985, **23**, 835, 839; (e) S. Takano, S. Sato, E. Goto, and K. Ogasawara, *J. Chem. Soc., Chem. Commun.*, 1986, 156.

2 F. Ungemach and J. M. Cook, *Heterocycles*, 1978, **9**, 1089.

3 A. H. Jackson, B. Naidoo, and P. Smith, *Tetrahedron*, 1968, **24**, 6119.

4 J. R. Frost, B. R. P. Gaudilliere, and A. E. Wick, *J. Chem. Soc., Chem. Commun.*, 1985, 895.

5 G. Casnati, A. Dossena, and A. Pochini, *Tetrahedron Lett.*, 1972, **13**, 5277.

6 F. Ungemach, M. DiPierro, R. Weber, and J. M. Cook, *J. Org. Chem.*, 1981, **46**, 164.

7 J. E. Baldwin, *J. Chem. Soc., Chem. Commun.*, 1976, 734.

8 C. Schopf and J. Thesing, *Angew. Chem.*, 1951, **63**, 377 (*Chem. Abstr.*, 1954, **48**, 4509e).

9 *cf.* J. Thesing, *Chem. Ber.*, 1954, **87**, 507 (*Chem. Abstr.*, 1955, **49** 9616f).

10 W. C. Still, M. Kahn, and A. Mitra, *J. Org. Chem.*, 1978, **43**, 2923.

11 E. J. Corey and J. W. Suggs, *Tetrahedron Lett.*, 1975, **16**, 2647.

Scientific paper workshop

The aim of this exercise is for students to look at a recently published scientific paper, use their chemistry knowledge to anticipate the steps the researcher might take, and to come up with solutions to problems the author encountered. It is important to demonstrate that the results presented in a paper are the final outcome of numerous ideas, and that undergraduate students are capable of coming up with good, viable solutions to research problems, even though the paper only reports one approach to the particular problem.

Reasonable answers to the questions in the student handouts are provided in Appendix A.

The exercise works best if no preparatory work is expected of the students – it loses some impact if the students have read a lot of background information. Some background information should be given on the first student handout in order to set the scene and outline the task.

Approximately two hours of total contact time are required to run this exercise, although a variety of slots can be used; for example, a single two hour workshop, three fifty minute sessions (each one tackling two sections of the exercise with slightly more discussion possible), or two fifty minute slots with some private study in between. The timing can be manipulated either to extend the exercise, by encouraging more discussion, or to shorten the exercise, by bringing the plenary sessions to a rapid conclusion.

The workshop is not restricted to groups (students could work as individuals or in pairs), although the discussion sessions seem to work particularly well with this arrangement.

This exercise is quite labour intensive on the part of the tutor, who will need to:

■ introduce the paper;

■ ensure the exercise keeps to time;

■ circulate, to ensure discussions are progressing well [additional tutor(s) might help here];

■ prompt students for answers in plenary sessions; and

■ summarise conclusions at the end of the plenary sessions.

Some ideas for papers to use in this exercise are given below. A number have been broken down in the suggested manner, and in one case questions and answers are also provided for the tutor.

Designing similar exercises

The following notes are given to guide the development of a new exercise. Suggested sources are letters, communications, or short papers, which can be split into sections which are discussed sequentially. The guidelines below assume that the students do not see the complete paper until the end of the exercise.

Most papers can be split into the following sections:

- Introduction

- Experimental section

- Results

- Conclusions

The content of the paper should be divided into several parts and provided on separate handouts. It can also be edited to reduce the amount of information the students have to deal with. Each of these sections can then be studied in turn, and the students' understanding/interpretation confirmed with questions.

Introduction
From the introduction the students should understand why the work described was important. In some instances background information may also be required in order to explain unfamiliar concepts and place the paper's contents in context. If a key statement is made in the introduction, the students could be asked to discuss this. Questions could be structured in a way that guides students to the key points. They could also discuss the choice of any measurement techniques or experimental methods, perhaps suggesting alternatives.

Experimental Section
If possible, students should be asked to draw on their experience in the laboratory when considering the experiments in the paper, and they should decide how they would have planned a similar study. Potential hazards or difficulties and how they should be dealt with could also be covered.

Results
The results section often needs to be split into several sub-sections, so that students are supplied with the 'raw' data if possible, and are asked to consider how it should be interpreted; for example, if spectral data are included, the structure of the sample could be determined. In the case of physical chemistry, the students could be prompted to draw out key features in the data, calculate constants and explain trends. Methods used to present data, such as tables, graphs or spectra, could be discussed and which is most appropriate for a particular situation decided. Students might then discuss the authors' interpretation of the results.

Conclusion
Students could be asked to provide their own conclusions based on the introduction to the paper, and the discussion of results, which could be compared to those of the authors. They could also consider further research that could be undertaken as a result of the paper, or suggest how the results might be of use in 'real life'.

Other suitable papers

The following papers from *Journal of the Chemical Society, Chemical Communications* between 1995–97, exemplify a range of suitable articles. Three of these articles have been further divided into sections and expanded.

1. A saccharide sponge. Synthesis and properties of dendritic boronic acid. T. D. James, H. Shinmori, M. Takeuchi & S. Shinkai, *J. Chem. Soc., Chem. Commun.*, 1996, 705

This paper describes how dendrimers that selectively bind to saccharides via boronic acid groups can be prepared.

i   Background
Why might the detection of saccharides be useful? How might you detect them? What are dendrimers? Why might they be useful as sensors?

ii   Background experiments
What features of **2** are crucial to its use as a sensor? From Figure 2, what is the binding constant for D-fructuose to **2**?

iii   Experimental Section (Part A)
Why did the authors not simply use anthracenylmethyl bromide in the first step of Scheme 1? What are the likely practical problems in making a dendrimer?

iv   Experimental Section (Part B)
What is the binding constant for D-fructose to dendrimer **1**? Are there likely to be any complications with the binding? Is this reflected in the binding data?

v   Conclusions
What conclusions do you draw? Is the dendrimer useful?

---

2. New sensor for dissolved dioxygen: a gold electrode modified with a condensation polymer film of β-cylcodextrin hosting cobalt tetraphenylporphyrin. F. D'Souza, Y.-Y. Hsieh, H. Wickman & W. Kutner, *J. Chem. Soc., Chem. Commun.*, 1997, 1191–1192.

i   Background
Why might a sensor by useful? How might you detect $[O_2]$ in solution? What is a cyclodextrin?

ii   Introduction
Explain how the porphyrin, cyclodextrin, polymer and gold film are all intended to link together to form a sensor. What advantages might the sensor in the paper have over other methods for analysing $[O_2]$?

iii   Preparation
Suggest how you might practically carry out the preparation of the new sensor. Did the authors use your method?

iv   Results
Is the peak current linearly correlated with the $[O_2]$? At what potential would you monitor in order to get the most accurate measurement of $[O_2]$?

v   Conclusions
Do you think the system could be used as a sensor for dioxygen? What problems might there be with it, compared with other oxygen sensors that are available?

3. First example of a copper(I)-water bond. Synthesis and structure of polymeric poly-μ-2,3-diphenylquinoxaline-(aqua)copper(I) cation. J. P. Naskar, S. Hati, D. Dutta & D. A. Tocher, *J. Chem. Soc., Chem. Commun.*, 1997, 1319.

i Background
Why is Cu(I) unstable in water? From the K value, how long would it take for the [Cu$^+$] to halve in aqueous solution? How might you try and prepare a stable Cu(I)-water complex?

ii Background experiments
How would you prepare ligands L^1, L^2, and L^3?

iii Predictions
What characteristics would you predict for **1** concerning (a) its magnetic properties; (b) its colour; and (c) its stability to oxidation?

iv Results
From Figure 1(a) where are the Cu(1) and H$_2$O moieties? How much of the π-system in L^3 is planar?

v Conclusions
Does the close up of the X-ray structure confirm your initial interpretation? Why might the preparation of a Cu(I)-water complex be useful or interesting?

- Formation of HNCO during catalytic reduction of NO$_x$ with olefins over Cu/ZSM-5. F. Radthe, R. A. Koeppel & A. Barker, *J. Chem. Soc., Chem. Commun.*, 1995, 427–428.

- Facile formation of a *cis*-platin-nonapeptide complex of human DNA polymerase-α origin. R. N. Bose, D. Li, M. Kennedy & S. Basu, *J. Chem. Soc., Chem. Commun.*, 1995, 1731–1732.

- An electrostatic investigation: how polar are ionic surfactant hydrocarbon tails? S. R. Gadre & S. S. Pingale, *J. Chem. Soc., Chem. Commun.*, 1996, 595–596.

- It's on lithium! An answer to the recent communication which asked the question: 'if the cyano ligand is not on copper, then where is it?' S. H. Bertz, G. Miao & M. Eriksson, *J. Chem. Soc., Chem. Commun.*, 1996, 815–816.

- Reversible dissolution/deposition of gold in iodine-iodine-acetonitrile systems. Y. Nakao & K. Sone, *J. Chem. Soc., Chem. Commun.*, 1996, 897–898.

- Hypervalent iodine-induced oxidative nucleophilic additions to alkenes: a novel acetoxy thiocyanation reaction in 1,1,1,3,3,3-hexafluoropropan-2-ol. A. De Mico, R. Margarita, A. Mariani & G. Piancatelli, *J. Chem. Soc., Chem. Commun.*, 1997, 1237–1238.

- *In situ* probing of surface sites on supported molybdenum nitride catalyst by CO adsorption. S. Yang, C. Li, J. Xu & Q. Xin, *J. Chem. Soc., Chem. Commun.*, 1997, 1247–1248.

- A pH cleavable linker for zone diffusion assays and single bead solution screens in combinatorial chemistry. B. Atrash & M. Bradley, *J. Chem. Soc., Chem. Commun.*, 1997, 1397–1398.

- Probing peristatic chirality of alkaline cations: NMR study of alkaline borocryptates. E. Graff, R. Graff, M. Wais Hasseini, C. Huguenard & F. Taulelle, *J. Chem. Soc., Chem. Commun.*, 1997, 1459–1460.

- Self-replication in a Diels-Alder reaction. B. Wang & I. O. Sutherland, *J. Chem. Soc., Chem. Commun.*, 1997, 1495–1496.

- Immobilization and cleavage of DNA at cationic, self-assembled monolayers containing C_{60} on gold. N. Higashi, T. Inoue & M. Niwa, *J. Chem. Soc., Chem. Commun.*, 1997, 1507–1508.

Assessment

The fifty-word abstract required in Section 6 of this exercise, submitted by groups or individuals, can be assessed. Alternative assessment exercises include asking students to:

- referee the paper using guidelines from a journal;

- describe an alternative means of studying the same research problem;

- outline a research proposal that might follow on from the results of the paper; or

- describe a potential application of the results.

3. Computer Keyboard Skills

Summary

Outline of the exercise

In this exercise students participate in the production of material for a new World Wide Web-based undergraduate 'book'. They are required to include text and a variety of chemistry graphics in a specific format. A number of different scenarios are outlined in the tutor's guide including one in which the preparation of structures and brief text on four natural products, and an outline for the mechanism of the Wittig reaction, are required. The aim of the exercise is to develop the ability to use chemical drawing packages and prepare text and graphics in an effective manner and it therefore provides a useful introduction to some of the most important computer keyboard skills required by chemists. The notes assume that students have access to appropriate computing facilities.

Key aims

■ to practise using standard computer drawing and word-processing packages for the production of effective visual aids

Time requirements

■ 1 hour introduction (tutor contact time)

■ 9 hours private study

■ 10 hours total student time

Timetable

The following timetable is suggested:

1 hour	Introduction	(possibly in small groups, in a computer room)
9 hours	Preparation	(private study)

Computer keyboard skills

A publisher is putting together a new undergraduate chemistry course on the World Wide Web, and is collecting samples of possible material before commissioning the whole course. You are an undergraduate chemist – you know the types of format and style which catch your attention, and make chemistry clear to you, so you are in an ideal position to win the deal for us! In this exercise you must construct two sample pages for the new course.

1. Natural Products

Prepare one page showing four natural products. Display their structures and describe some of their chemistry and properties. An example of chlorophyll a is shown below, to illustrate the layout required:

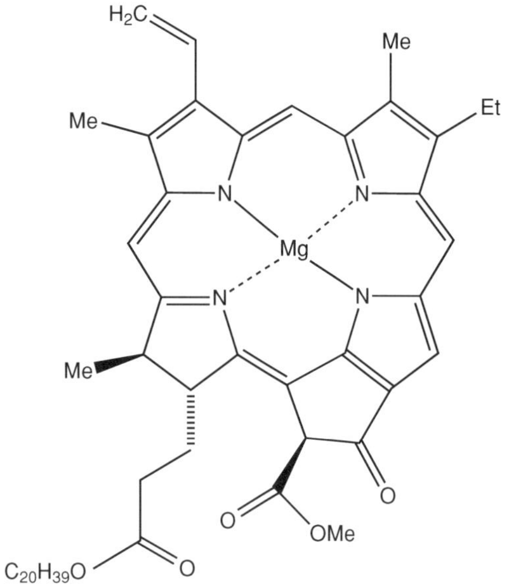

Chlorophyll a

■ A green pigment found in most plants

■ One of several natural products containing a porphyrin ring (four pyrroles connected by CH groups)

■ A catalyst for photosynthesis

■ First total synthesis by Woodward

Prepare the following four structures:

■ Cholesterol (using a 3D representation to show ring conformations)

■ Strychnine

■ Quinine

■ Another interesting natural product, of similar complexity, of your choice.

The sample must fit on one side of A4, and must use text with added graphics (so that hyperlinks can be added for the final publication). A 'table' format can be used to achieve this.

2. Wittig Reaction

Prepare a one-page summary of the Wittig reaction of a stabilised ylid with an aldehyde, clearly explaining the stereochemistry. The sample must fit on one side of A4, and can be constructed entirely in a chemical drawing package. The summary should contain:

i) A detailed mechanism for the reaction of $Ph_3P=CHCO_2Et$ with Ph–CHO, showing the intermediates and stereochemistry leading to **E-** and **Z-** products.

ii) A verbal description of the main features of the reaction mechanism and comments on the stereoselectivity – this can be in the form of annotations of the mechanism.

iii) A diagram of the reaction profile (reaction co-ordinate versus energy), including the key intermediates in the reaction sequence, namely the betaine and the oxaphosphetane, expected for this particular Wittig reaction.

Specific formats are required for both pages. Times 12 point font should be used for most of the text, although headings can be larger. The following settings must be used within the chemical drawing package:

Drawing settings

- Chain angle 120°
- Bond spacing 12 %
- Fixed length 0.773 cm
- Bold width 0.902 cm
- Line width 0.026 cm
- Margin width 0.053 cm
- Hash spacing 0.071 cm

Text settings

Caption Text Settings
Helvetica 8 pt Normal

Label Text Settings
Helvetica 10 pt Bold
Fractional character width x

Preferences

Units cm

Tolerance 5 pixels

Remember – computers, disks and printers often go wrong at the last minute. Prepare your submissions in good time, prepare back-up copies of your work every 5 minutes, and make sure it is printed well before the deadline.

T3 **Computer keyboard skills**

Many students will have already received training in word-processing, and guidance on the use of a range of other computing facilities, for example email, the World Wide Web (WWW), databases and spreadsheets. The simple format of this short exercise requires them to develop several skills that will be particularly useful in other aspects of communicating chemistry including:

- confidence with a basic chemical drawing package;

- a feel for the range of chemical drawing graphics available;

- mixing text and graphics; and

- following a precise format.

A one hour session can be used to introduce students to word-processing and chemical drawing packages, although the content of the session will depend on students' previous experience and the specific hardware and software available. It is effective to carry out this session in a computer room. This exercise provides a useful introduction to some of the most important keyboard skills required by chemists.

The Exercises

Examples of two exercises concerning Natural Products and the Wittig Reaction are given in the student's guide. Model answers for these exercises are given below. These exercises are specific to organic chemistry, however similar exercises in inorganic or physical chemistry can be put together easily. For example:

a Draw the structures of four important inorganic molecules and describe some of their chemistry and properties. For example:
- Cp-Fe(CO)(PPh$_3$)(COMe) – what is it used for?
  See *Chem. Br.*, 1989, **25(3)**, 268.

- 3-(η–Cp)-3,1,2,-*closo*-CoC$_2$B$_9$H$_{11}$ – discuss its isomerisation.
  See *J. Am. Chem. Soc.*, 1972, **94**, 6679.

- XeF$_6$ – what is its shape, and why was this unexpected?

- *cis*-PtCl$_2$(NH$_3$)$_2$ – what is its medicinal use, and how does it work?

b Discuss the mechanism by which Mn(CO)$_5$Me isomerises to Mn(CO)$_4$(COMe). Clearly show the mechanism for the reaction, explain the key experiments that aided the elucidation of the mechanism and draw a reaction co-ordinate/energy profile diagram; also indicate why the elucidation of this mechanism was important for other organometallic processes.

c Explain four important physical chemistry terms; each explanation should include an equation and a graphic (eg 'Boltzmann distribution', 'first order kinetics', 'the Arrhenius equation', 'diatomic dissociation energy').

d Explain how a specific piece of equipment can be used to determine a key physical constant (*eg* bond length from infrared spectroscopy, second order rate constant measurements using conductivity apparatus, heat of evaporation of a compound from vapour pressure measurements).

The explanation should include a schematic drawing of the equipment, a description of the experiment, a graphical representation of the results, and how these would be interpreted.

Adapting the exercise

It is easy to adapt this exercise to specific course requirements. One variant is to have a pool of about twenty natural products, and about six reactions, from which the students are randomly allocated their task. This reduces the risk of copying and adds a little more interest and individuality to the exercise.

Assessment

It is simple to mark the submissions with a grade (A: excellent, B: good; C: average; D: poor; E: very poor; X: unsatisfactory), and peer-assessment could easily be used.

Model Answers

Cholesterol
- Polycyclic, hydrophobic hydrocarbon with a single hydrophilic OH group
- Constituent of cell walls
- Biosynthesised from C_5 building blocks
- Can be deposited on the walls of arteries, leading to heart disease

Strychnine
- Alkaloid (naturally occurring base)
- Extremely toxic
- First synthesis by Woodward, but Overman has recently completed the first enantiospecific synthesis

Quinine
- Alkaloid isolated from Cinchona tree
- Structure determined in 1908
- Used to treat malaria
- First synthesis by Woodward in 1944

Penicillin G
- The penicillins are fungal antibiotics
- They all contain the strained 4-membered β-lactam ring, and an additional 5-membered N/S ring
- They work by inhibiting a key enzyme involved in the construction of bacterial cell walls
- Penicillins with various R groups are used medicinally, of which penicillin G (R = CH_2Ph) has been used extensively.

Pencillin G: R = CH_2Ph

The Wittig reaction between benzaldehyde and the phosphorus ylid of ethyl bromoethanoate

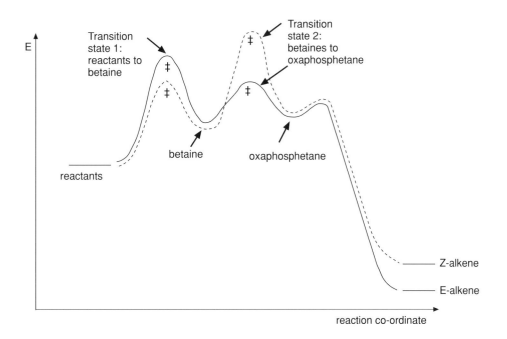

1. Initial attack of the ylid on the C=O is as shown, forming **A** (Ph$_3$P and O *anti*, and Ph and CO$_2$Et *anti*.).

2. But the first step is reversible, so the betaines **A** and **B** can both be formed.

3. The betaines are converted into the oxaphosphetanes **C** and **D**. **D** forms much faster.

4. Intermediate **D** rapidly collapses to *E*-alkene, and the initial equilibria allow almost all of the reaction to proceed *via* **B** and **D**.

The following diagram illustrates the energy versus reaction co-ordinate profiles for the two reaction pathways.

4. World Wide Web Treasure Hunt

Summary

Outline of the exercise

Students undertake a 'treasure hunt' to locate a variety of items on the World Wide Web (WWW). A wealth of useful chemistry information is available on the WWW covering exciting new developments, companies, university departments, people, events, literature, databases, safety, and basic chemistry and this exercise is primarily intended to be a way for students to develop their ability to locate some of this material. Although some students are confident at using computers and the web this exercise should help to give weaker students confidence.

Key aims

■ to introduce information retrieval on the WWW; and

■ to develop problem solving skills.

Time requirements

■ 1 hour introduction (tutor contact time)

■ 7 hours private study

■ 8 hours total student time

Timetable

The following timetable is suggested:

1 hour	Introduction	(lecture slot)
7 hours	Preparation	(private study, at a computer)

S4 World wide web treasure hunt

Treasure Hunt 1

This exercise aims to

■ introduce you to the World Wide Web;

■ show you how much chemistry information is available on the internet; and

■ develop your information retrieval skills.

Find the answers to the following questions by locating the relevant material from the Web. Submit both the answers and the URLs indicating where the information can be found.

1. List three areas of chemical research which are undertaken in this department.

2. What is the electron configuration of iron?

3. Name three Royal Society of Chemistry journals in the area of analytical chemistry which can be accessed on-line.

4. Name four research institutes sponsored by the Biotechnology and Biological Science Research Council (BBSRC).

5. Who won this year's Nobel Prize for Chemistry?

6. What is the postal address of the Royal Society?

7. Suggest a URL which might be used in next year's treasure hunt (chemistry and science URLs are preferred, but imaginative educational suggestions are also welcome).

Treasure Hunt 2

This exercise aims to

■ introduce you to the World Wide Web;

■ show you how much chemistry information is available on the internet; and

■ develop your information retrieval skills.

Find the answers to the following questions by locating the relevant material from the WWW. Submit both the answers and the URLs indicating where the information can be found.

1. List three areas of chemical research which are undertaken in this department.

2. What are the hazards of oxygen?

3. Name three Royal Society of Chemistry journals in the area of organic chemistry which can be accessed on-line.

4. What are the eight main areas of science funded by the Engineering and Physical Sciences Research Council (EPSRC)?

5. What was Bristol University's Molecule of the Month for April 1997?

6. Which transition element is studied in the Virtual Laboratory at Oxford University?

7. Suggest a URL which might be used in next year's treasure hunt (chemistry and science URLs are preferred, but imaginative educational suggestions are also welcome).

Treasure Hunt 3

This exercise aims to

■ introduce you to the World Wide Web;

■ show you how much chemistry information is available on the internet; and

■ develop your information retrieval skills.

Find the answers to the following questions by locating the relevant material from the WWW. Submit both the answers and the URLs indicating where the information can be found.

1. List three areas of chemical research which are undertaken in this department.

2. What are the mineral sources of tantalum?

3. Name three Royal Society of Chemistry journals in the area of physical chemistry which can be accessed on-line.

4. How many studentships does the Medical Research Council (MRC) fund annually?

5. What was Bristol University's Molecule of the Month for August 1997?

6. What is the postal address of the Royal Society?

7. Suggest a URL which might be used in next year's treasure hunt (chemistry and science URLs are preferred, but imaginative educational suggestions are also welcome).

Treasure Hunt 4

This exercise aims to

■ introduce you to the World Wide Web;

■ show you how much chemistry information is available on the internet; and

■ develop your information retrieval skills.

Find the answers to the following questions by locating the relevant material from the Web. Submit both the answers and the URLs indicating where the information can be found.

1. List three areas of chemical research undertaken in this department.

2. What is the element osmium used for?

3. Name two Royal Society of Chemistry journals in the area of inorganic and material chemistry which can be accessed on-line.

4. Name the four main areas of research which are funded by the Natural Environmental Research Council (NERC).

5. Who won this year's Nobel Prize for Chemistry?

6. What is the Research Assessment Exercise (RAE) grade for this university's chemistry department?

7. Suggest a URL which might be used in next year's treasure hunt (chemistry and science URLs are preferred, but imaginative educational suggestions are also welcome).

Treasure Hunt 5

This exercise aims to

■ introduce you to the World Wide Web;

■ show you how much chemistry information is available on the internet; and

■ develop your information retrieval skills.

Find the answers to the following questions by locating the relevant material from the Web. Submit both the answers and the URLs indicating where the information can be found.

1. List three areas of chemical research undertaken in this department.

2. What is the calculated lattice energy of KH?

3. Name three general Royal Society of Chemistry journals which can be accessed on-line.

4. Name the three sites supported by the Council for the Central Laboratories of the Research Councils (CCLRC).

5. Who won this year's Nobel Prize for Chemistry?

6. Which transition element is studied in the Virtual Laboratory at Oxford University?

7. Suggest a URL which might be used in next year's treasure hunt (chemistry and science URLs are preferred, but imaginative educational suggestions are also welcome).

Treasure Hunt 6

This exercise aims to

■ introduce you to the World Wide Web;

■ show you how much chemistry information is available on the internet; and

■ develop your information retrieval skills.

Find the answers to the following questions by locating the relevant material from the WWW. Submit both the answers and the URLs indicating where the information can be found.

1. List three areas of chemical research undertaken in this department.

2. What is the abundance of gold in the Earth's crust?

3. Give the URL for Bath Information and Data Services homepage.

4. What is the name of the careers handbook published by the Royal Society of Chemistry for undergraduate chemistry students?

5. What is the postal address of the Royal Society?

6. Which transition element is studied in the Virtual Laboratory at Oxford University?

7. Suggest a URL which might be used in next year's treasure hunt (chemistry and science URLs are preferred, but imaginative educational suggestions are also welcome).

Treasure Hunt 7

This exercise aims to

■ introduce you to the World Wide Web;

■ show you how much chemistry information is available on the internet; and

■ develop your information retrieval skills.

Find the answers to the following questions by locating the relevant material from the WWW. Submit both the answers and the URLs indicating where the information can be found.

1. List three areas of chemical research undertaken in this department.

2. Name a radioisotope of vanadium and give its half-life.

3. List three databases or databanks containing chemical information which are available from the Royal Society of Chemistry.

4. Give the URL of the jobs database supported by the New Scientist magazine.

5. Which transition element is studied in the Virtual Laboratory at Oxford University?

6. What is the Research Assessment Exercise (RAE) grade of this university's chemistry department?

7. Suggest a URL which might be used in next year's treasure hunt (chemistry and science URLs are preferred, but imaginative educational suggestions are also welcome).

Treasure Hunt 8

This exercise aims to

■ introduce you to the World Wide Web;

■ show you how much chemistry information is available on the internet; and

■ develop your information retrieval skills.

Find the answers to the following questions by locating the relevant material from the WWW. Submit both the answers and the URLs indicating where the information can be found.

1. List three areas of chemical research undertaken in this department.

2. Give an example of a compound which contains xenon with oxidation number 6.

3. Which UK university hosts the Beilstein database? Give its URL.

4. The Royal Society of Chemistry provides careers advice on the internet. What suggestions are given for preparation for a job interview?

5. What was Bristol University's Molecule of the Month for May 1998?

6. What is the Research Assessment Exercise (RAE) grade of this university's chemistry department?

7. Suggest a URL which might be used in next year's treasure hunt (chemistry and science URLs are preferred, but imaginative educational suggestions are also welcome).

Treasure Hunt 9

This exercise aims to

■ introduce you to the World Wide Web;

■ show you how much chemistry information is available on the internet; and

■ develop your information retrieval skills.

Find the answers to the following questions by locating the relevant material from the WWW. Submit both the answers and the URLs indicating where the information can be found.

1. List three areas of chemical research undertaken in this department.

2. What is the chemical symbol of element 107?

3. How many structures can be found in the Cambridge Structural Database?

4. Give the URL of the Royal Society of Chemistry's job advertisement website.

5. What is the postal address of the Royal Society?

6. What is the URL of a guide to drug metabolism provided by a leading pharmaceutical company?

7. Suggest a URL which might be used in next year's treasure hunt (chemistry and science URLs are preferred, but imaginative educational suggestions are also welcome).

Treasure Hunt 10

This exercise aims to

■ introduce you to the World Wide Web;

■ show you how much chemistry information is available on the internet; and

■ develop your information retrieval skills.

Find the answers to the following questions by locating the relevant material from the WWW. Submit both the answers and the URLs indicating where the information can be found.

1. List three areas of chemical research undertaken in this department.

2. What is the biological role of arsenic?

3. Where can you obtain information about Chemical Abstracts on the WWW?

4. What is the jobseeker service offered by the Royal Society of Chemistry?

5. Who won this year's Nobel Prize for Chemistry?

6. What is the Research Assessment Exercise (RAE) grade of this university's chemistry department?

7. Suggest a URL which might be used in next year's treasure hunt (chemistry and science URLs are preferred, but imaginative educational suggestions are also welcome).

World wide web treasure hunt

This exercise is a useful short introduction to the web; although some students are very confident using the computer (and will complete the treasure hunt in a couple of hours), this exercise is sufficient to give the weaker students confidence. Students seem to find the exercise enjoyable, and discover the extent to which useful chemistry information is available on the web.

The students are required to locate seven pieces of 'treasure', and must submit their results to the tutor before a deadline.

The exercise can be run using hard copies of the treasure hunt. Ten examples are provided in this book and can be randomly allocated to students. Students might be required to submit their results (ie the answers and the URLs where the answers to the questions can be found) to the tutor via email. It is also possible to run the exercise directly from the web by producing tailor-made pages.

The introduction (nominally scheduled for one hour) can be run in a number of ways, influenced by the experience of the students, the size of the group, and the facilities available. However, it is more important to let students 'have a go', than to spend a lot of time explaining the power of the web. A five-minute introduction to the web, and the nature of the exercise may be adequate. If a more in-depth introduction is required, performing a search in order to answer one of the questions is effective.

Answers

The URLs below give, as far as possible, the exact location where required information can be found, and in some cases an indication of how to reach the website. All the URLs listed were correct at the time of printing. However, URLs are often subject to change and the authors apologise for any inconvenience that this may cause.

Within each treasure hunt the following types of question are asked.

Type 1 questions
These questions relate to information about the research carried out by a particular chemistry department.

Type 2 questions
These questions relate to individual elements, their properites or their reactivity. Their answers can all be found via the University of Sheffield's Web Elements URL – http://www.sheffield.ac.uk/chemistry/web_elements followed by the symbol of the element concerned (ie Fe, O, Ta ...)

What is the electron configuration of iron?
$[Ar]3d^6 4s^2$

What are the hazards of oxygen?
At high partial pressures, oxygen can cause convulsions, pulmonary changes and teratogenic effects. Ozone, peroxide, and superoxide are highly toxic. Too little oxygen results in asphyxiation. Oxygen-enriched air is a fire hazard because the burning rate of combustible materials is increased.

What are the mineral sources of tantalum?
Stony and iron meteorites and the earth's crust

What is the chemical symbol of element 107?
Bh

What is the abundance of gold in the Earth's crust (in ppm)?
0.002 ppm

Name a radioisotope of vanadium and give its half-life.

radioisotope	half-life
^{47}V	32.6 mins
^{48}V	15.98 days
^{49}V	337 days
^{52}V	3.76 mins
^{53}V	1.61 mins

What is the calculated lattice energy of KH?
699 kJ mol^{-1}

What is the element osmium used for?
In combination with other metals of the platinum group, osmium is often used to produce very hard alloys. These are used for fountain pen nibs, instrument pivots and electrical contacts. Osmium tetraoxide is used in forensic science to detect fingerprints. The platinum/osmium 90/10 alloy is used in medical implants such as pacemakers and replacement valves.

Give an example of a compound which contains xenon with oxidation number 6.
XeF_6

What is the biological role of arsenic?
Arsenic is a necessary ultratrace element for a number of species including red algae, chickens, rats, goats and pigs. It may also be necessary for humans. Deficiency results in inhibited growth.

Type 3 questions
The following questions all relate to information retrieval on the web.

All on-line RSC journals can be accessed from
http://www.rsc.org/is/journals/current/ejs.htm:

■ *Analytical chemistry*
Analyst, Analytical Communications and Journal of Analytical Atomic Spectroscopy

■ *Organic chemistry*
Natural Product Reports and Perkin Transactions 1 and 2

■ *Physical chemistry*
Faraday Transactions and Faraday Discussions

■ *Inorganic and materials chemistry*
Dalton Transactions and Journal of Materials Chemistry

■ *General*
Annual reports on the Progress of Chemistry, Chemical Communications, New Journal of Chemistry, Mendeleev Communications, Journal of Chemical Research and Russian Chemical Reviews

Give the URL of the Bath Information and Data Services homepage.
http://www.bids.ac.uk

List three databases or databanks containing chemical information which are available from the Royal Society of Chemistry.
Agrochemicals; Analytical chemistry; Biotechnology; Chemical business; Chemical engineering; Health, safety & environment; Nutrition; Organic chemistry; Natural Products Updates
http://www.rsc.org/is/database/sec_serv.htm

Which university hosts the Beilstein database?
Manchester University
http://midas.ac.uk/crossfire

How many structures can be found in the Cambridge Structural Database?
More than 180 000
http://www.ccdc.cam.ac.uk/prods/csd.html

Where can you obtain information about the Chemical Abstracts Service on the web?
http://www.cas.org

Type 4 questions
The following questions relate to careers information and information about research councils.

Name four research institutes sponsored by the Biotechnology and Biological Science Research Council (BBSRC).
These are Babraham Institute, Institute of Arable Crops Research, Institute for Animal Health, Institute of Food Research, Institute of Grassland and Environmental Research, John Innes Centre, Roslin Institute, and Silsoe Research Institute.
http://www.bbsrc.ac.uk/

Name the four main areas of research which are funded by the Natural Environmental Research Council (NERC).
Atmospheric Science and Technology, Earth Science and Technology, Marine Science and Technology; Terrestrial & Freshwater Science and Technology
http://www.nerc.ac.uk/cgibin/loadframeset?title=Science&right=/science/&left=/main-menu.html

What are the eight main areas of science funded by the Engineering and Physical Sciences Research Council (EPSRC)?
Chemistry, Physics, Mathematics, General Engineering, Engineering for the environment, infrastructure and healthcare, Engineering for manufacture, Information Technology and computer science and Materials
http://www.epsrc.ac.uk/progs/progfram.htm

Name the three sites supported by the Council for the Central Laboratories of the Research Councils (CCLRC).
Daresbury Laboratory in Cheshire, Rutherford Appleton Laboratory in Oxfordshire, and Chilbolton Observatory in Hampshire
http://www.cclrc.ac.uk

How many studentships does the Medical Research Council (MRC) fund annually?
Approximately 1200 studentships are sponsored annually.
http://www.mrc.ac.uk/left_5b.html

What is the name of the careers handbook published by the Royal Society of Chemistry for undergraduate chemistry students?
Got a Degree...What Next?
http://www.rsc.org/lap/educatio/highered.htm

Give the URL of the jobs database supported by the New Scientist magazine.
http://www.sciencejobs.com which can be accessed via
http://www.newscientist.com

Give the URL of the Royal Society of Chemistry's job advertisement website.
http://www.chemsoc.org/gateway/chembyte/careers.htm

What is the jobseeker service offered by the Royal Society of Chemistry?
The Society offers a free 'Jobseeker' service to all members needing advice
on employment issues. Jobseeker services include:

■ CV advisory service;

■ employment placing service in partnership with a specialist scientific
recruitment agency;

■ opportunities to participate in workshop and training activities on topics
including CV writing, interview techniques and assertiveness;

■ reference publications on employment related subjects, including a
Guide to Consultancy;

■ confidential telephone advice line.

http://www.rsc.org/lap/profserv/profserv.htm

The Royal Society of Chemistry provides careers advice on the internet.
What suggestions are given for preparation for a job interview?
You should find out as much as you can about the organisation. Companies
are usually very helpful in sending out their annual reports on request and it
is also worth checking to see if they have a World Wide Web site (the
Kompass directory should help you). This should give you an overall view of
the type of organisation to which you are applying and may give you some
ideas for possible interview questions.

You should find out:

■ what the company or organisation does

■ what the company produces

■ the company's origins

■ where the company operates

■ the type and number of company employees

■ current initiatives and new products

■ who the head of the organisation is

■ any recent press comment.

http://www.chemsoc.org/gateway/careersbefore.htm

Types 5 and 6 questions
These questions are made up from a selection of the following miscellaneous
questions.

What is the postal address of the Royal Society?
6 Carlton House Terrace
LONDON SW1Y 5AG
http://www.royalsoc.ac.uk/

What is the Research Assessment Exercise (RAE) grade for this university's
chemistry department?
These can all be obtained via http://www.niss.ac.uk/education/hefc/rae96/,
or will be directly obtainable via a specific department's own website.

Who won this year's Nobel Prize for Chemistry?
http://www.nobel.se/laureates/chemistry-YEAR.html

What is Bristol University's Molecule of the Month for February/April 1997 and May 1998?
This can be accessed via the RSC's website at
http://www.chemsoc.org/links/toptwenty.htm
Its actual URL is http://www.bris.ac.uk/Depts/Chemistry/MOTM/motm.htm
February 1997 – Sulfanilamide
April 1997 – Cyclooctene
May 1998 - Proline

Which transition element is studied in the Virtual Laboratory at Oxford University?
Ni or Fe
http://neon.chem.ox.ac.uk/vrchemistry/

What is the URL of a guide to drug metabolism provided by a leading pharmaceutical company?
http://www.glaxowellcome.co.uk/science/drugmet

Assessment

The students must hand in the URLs of their seven 'treasures' as well as the answers to the questions – submission by e-mail is particularly appropriate. A mark can be given for the number of correct URLs, although they ought to be able to get full marks (especially if a tutor is available to offer help). However, the main point of the exercise is for students to explore the Internet and learn how to use it effectively to find chemical information.

5. New Chemist article

Summary

Outline of the exercise

Students work in pairs to produce a short article for a fictitious journal – the *New Chemist*. This journal only publishes articles relating to chemistry, which are written for a general audience of chemists at many levels (*ie* from professional chemists to those at secondary school), and its aim is to highlight recent developments in chemistry by identifying novel and exciting research reported in chemical literature. The style of the articles which students produce is prescribed in this exercise and is based on the half-page news articles found in the *New Scientist* magazine. Recent papers from *Chemical Communications* can be used as a basis for the articles. Some examples of work produced by students during trialling of the exercise are included in this book.

Key aims

■ To develop the skills of information retrieval (and comprehension), written delivery, and visual delivery.

Many students are poor at writing concise summaries, and at explaining chemistry to non-specialist audiences; these skills are required for this exercise.

Time requirements

■ Less than 1 hour (tutor contact time)

■ Approximately 18 hours private study

Timetable

Approximately 5 minutes are required to explain this exercise. Students need about 20 hours to to complete the exercise well – 9 hours for choosing and comprehending a suitable article, and 9 hours to write the article and format it correctly. A reasonably imminent deadline is helpful – if students have other course commitments, a timescale of 2 weeks is appropriate.

S5 New Chemist article

Publications such as *New Scientist*, and daily newspaper science pages keep us in touch with developments in chemistry and other subject areas; for example, with stories of 'Dolly' the cloned sheep, the Hale-Bopp comet, or the debate about the demise of the dinosaurs. These stories are usually picked up by a journalist with a training in science, who takes an interesting journal article (often a preliminary communication), and follows up the science to make a story. Some examples of chemistry articles in *New Scientist* are:

1. A. Coghlan, 'Water winds up the world's smallest spring' *New Scientist*, 1997, No. 2077, 16.

2. C. Seife, 'This way up – why the HIV drug AZT needs to stay in shape' *New Scientist*, 1998, No. 2129, 15.

3. D. Bradley, 'Ringing the changes: the hard slog of making steroids should soon be a distant memory' *New Scientist*, 1998, No. 2123, 14.

4. P. Hadfield & R. Walden, 'Catalysts for change: could some cunning chemistry make hydrogen the fuel of the 21st century?' *New Scientist*, 1998, No. 2123, 10.

In this exercise you and a colleague are required to produce a similar article for the *New Chemist* magazine. This publication reports recent advances in chemistry for a general audience of chemists from the professional to the secondary school pupil. Choose a paper that interests you from any recent issue of *Chemical Communications*. Your friends will be assessing the article you produce – so if they cannot understand it, non-specialists will have even greater difficulty! Instructions from the editor are given below and must be followed exactly. Some guidelines are also provided to help you.

Instructions and guidelines for authors

■ The article must be exactly one half side of A4;

■ It should follow the *New Scientist* format (font, font size, column width, etc). It is often easiest to write the text in an open A4 layout, and re-format it at the end;

■ It must contain at least one drawing/scheme/picture drawn with a chemical drawing package;

■ One additional graphic may be used (*eg* picture/graph/cartoon); and

■ It should be about 300 words in length (absolute limits 200–400 words).

■ Three copies of the finished article should be submitted before the deadline.

■ Choose a good title;

■ Use the first paragraph for a brief, catchy summary of what was achieved in the research;

■ Use the next three to four paragraphs to expand on the detail – remember that chemistry jargon will need to be explained to non-experts;

■ End with a punchy conclusion, possibly including an indication of how the chemistry might be developed or exploited in the future;

■ Ask a friendly expert (perhaps a chemistry lecturer or a fellow undergraduate) for comments; and

■ Make sure your names are on the article.

■ As soon as you and your colleague have agreed on a paper on which to base the article, write the reference on the sheet provided on the notice-board, and make sure that nobody else has already selected it. The deadline for submissions is also on the notice-board.

■ Disks corrupt, computer crash, and printers jam – the *New Chemist* deadline is rigorously enforced by the editor so do not leave production of the article until the last minute.

T5 New Chemist article

Provided the students have the appropriate computer keyboard skills, minimal tutor input is required for this exercise. The articles could be produced in many formats, but the style, level and length required here are ideal both for checking whether students understand the underlying chemistry and for developing their ability to explain chemistry to non-specialists in writing. This particular format also requires students to prepare computer-generated text including graphics.

There is real merit in having a very specific format for three reasons:

■ chemists need to write reports of a given format throughout their careers;

■ all the students produce the same length of submission; and

■ it is easier to compare and assess the articles.

Finally, the exercise runs best if students work in pairs, although it could also be carried out individually. Students benefit from the joint discussion and criticism of the article as it develops – a practice that is encouraged for all written reports, even if prepared by individuals. An element of peer pressure is also introduced when working in pairs, which tends to lead to the production of higher quality articles.

Designing similar exercises
It is very easy to choose different formats and levels for similar exercises. For example, students could convert some of their laboratory results into journal format. This works well if they have carried out a research project, although it is not realistic for normal laboratory work.

An attractive extension to the exercise is actually to produce a copy of *New Chemist* using students' articles. Students will know that the format must be exactly right, and that their colleagues will read their articles, so there is a strong incentive for them to do a good job.

Other *New Scientist* articles
The *New Scientist* articles listed below are all based on short reports found in *Chemical Communications*. They cover a range of chemical topics and could be used in addition to the examples given in the student's guide.

1. D. Bradley, 'Twists of life written in the stars' *New Scientist*, 1997, No. 2064, 17.

2. D. Bradley, 'Molecular impostor uncoils DNA theory' *New Scientist*, 1997, No. 2070, 19.

3. W. Wood, 'Glowing jellyfish just what the doctor ordered' *New Scientist*, 1997, No. 2071, 17.

4. D. Bradley, 'Buckyball 'transistor' raises nanocomputing hopes' *New Scientist*, 1997, No. 2072, 18.

5. D. Bradley, 'Catalyst banishes deadly mirror molecules' *New Scientist*, 1997, No. 2074, 18.

6. D. Bradley, 'Recipe for jelly promises new catalysts' *New Scientist*, 1997, No. 2076, 20.

7. A. Coghlan, 'Water winds up the world's smallest spring' *New Scientist*, 1997, No. 2077, 16.

8. C. Seife, 'This way up – why the HIV drug AZT needs to stay in shape' *New Scientist*, 1998, No. 2129, 15.

9. P. Hadfield & R. Walden, 'Catalysts for change: could some cunning chemistry make hydrogen the fuel of the 21st century' *New Scientist*, 1998, No. 2123, 10.

10. D. Bradley, 'Ringing the changes: the hard slog of making steroids should soon be a distant memory' *New Scientist*, 1998, No. 2123, 14.

Student examples

Two examples of *New Chemist* articles produced by students undertaking the exercise are included. These indicate the sort of articles that might reasonably be expected.

Assessment

A suggested marking scheme is included, and the tutor can carry out assessment. However, peer-assessment is also effective, with each pair of students allocated two articles to assess. Average marks can be awarded (if the two marks differ by less than 10%), with tutor input only necessary if the marks differ by more than 10%. Completed assessment forms should be returned to the students concerned, to provide feedback. The following are also worth considering:

a) An open discussion to agree a marking scheme (what criteria to use, and how many marks for each one); the students enjoy and benefit from actually thinking about how to assess their own work.

b) Asking the students to work in groups to assess several articles. The group discussion, and the chance to place a reasonable number of articles in rank order, produces more consistent marks.

Organic Chemists Redundant?

Dawn Robinson and Trish Drennan

MOLECULES that build themselves could lead to a new generation of molecular electronic devices and antibiotics. Such complicated systems could be constructed from molecules which could uniquely identify each other and self-assemble in a "jigsaw" like fashion.

Currently two classes of molecular building blocks are being used to construct primitive molecular switching devices and large molecular assemblies. These are namely the catenanes and rotaxanes, the former being two or more interlocking rings giving rise to a "chain" like structure and the latter being one or more rings threaded onto a dumbell shaped molecule where multiple threading gives rise to an "abacus" type molecule.

Catenane   Rotaxane

Synthesis of the "chain" type compounds is trivial with the [3]catenane, self-assembled in 25% yield by reacting a simple dibromide with a dicationic salt, in the presence of a templating agent, such as a

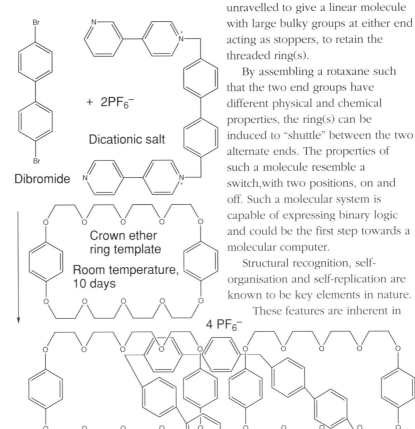

[3]catenane

crown ether ring. The reaction occurs at room temperature over a period of ten days (*Journal of the Chemical Society, Chemical Communications* 1996, No. 4, p.487).

Rotaxanes can be synthesised in a similar manner where a ring is unravelled to give a linear molecule with large bulky groups at either end acting as stoppers, to retain the threaded ring(s).

By assembling a rotaxane such that the two end groups have different physical and chemical properties, the ring(s) can be induced to "shuttle" between the two alternate ends. The properties of such a molecule resemble a switch, with two positions, on and off. Such a molecular system is capable of expressing binary logic and could be the first step towards a molecular computer.

Structural recognition, self-organisation and self-replication are known to be key elements in nature. These features are inherent in chemistry and subsequently could be used, with a bit of imagination, to mimic biological processes.

So perhaps the organic chemist is not quite redundant yet?!

Helping holes make computer screens better!

Neil Polwart and Martin Melia

COMPUTER screens on laptop computers may soon be able to have full colour displays if work done at Toyota's Research and Development Labs makes it to the production line.

Scientists believe that electro-luminescent (EL) devices made from layers of polymer are likely to hold the future for full colour flat panel display systems. Such displays would not only offer the full range of colours detectable to the human eye but would also need to use only low voltages, and be able to operate highly efficiently. Techniques are available to produce such devices, however they degrade very quickly. This instability, which results in a reduction in luminescence and an increase in drive voltage, is believed to be the result of changes in the arrangement of molecules in a thin film which carries electronic holes – vital to the properties of these devices.

Typically such devices use *N,N'-diphenyl-N,N'-(m-tolyl)benzidene* (TPD) as the hole carrier. As the device is used it heats up and reaches temperatures close to a critical point known as its glass transition temperature; here molecules move around in the thin film and this is believed to cause changes in their electrical properties.

Recently another group of scientists fabricated devices with long life times using a starburst shaped molecule (called TCTA); unfortunately the devices require almost three times as much energy to emit light than standard TPD devices.

TPD is made up from two smaller units called TPA; the team at Toyota's lab have managed to produce a series of compounds made from two to five of these units with increasing glass transition temperatures. EL devices have been fabricated with optical and electrical properties similar to those made from TPD, note the difference being that with these devices they could operate at 100 °C for 100 hours without serious damage, whilst the TPD device broke down after a few seconds at those temperatures. Toyota's team says in *Chemical Communications* (21/09/96, 18, 2175) this is directly linked to the glass transition temperatures.

Obviously some work still needs to be done before we see full colour EL devices in mass production, but perhaps in a few years time you will be reading an on-line version of *New Scientist* on a full colour

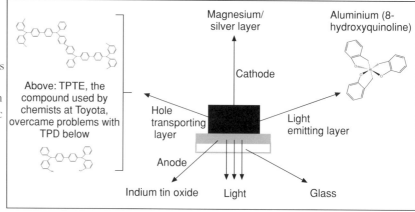

Above: TPTE, the compound used by chemists at Toyota, overcame problems with TPD below

Magnesium/silver layer

Aluminium (8-hydroxyquinoline)

Cathode

Hole transporting layer

Light emitting layer

Anode

Indium tin oxide   Light   Glass

Assessment form

Article title:

_____

Authors:

_____

Assessment criteria for scientific content and presentation

- Did the title/introduction make you want to read the article?   /10

- Was the article well written (ie sentences grammatically
  correct and easy to read)?   /10

- Was the article scientifically accurate?   /10

- Was the article written clearly and unambiguously?   /10

- Were the graphics clear, interesting, relevant, and informative?   /10

- Was the presentation and layout attractive?   /10

Total mark _____/60

Total mark as a percentage _____%

What was your overall impression of the quality of this report as a
percentage?

(Excellent, >80%; very good, 70-80%; good, 60-70%; average, 50-60%;
poor, 40-50%; very poor, <40%)

_____%

If appropriate, amend the marks given until the 'impression' mark awarded
matches the total percentage mark.

Final mark _____%

Brief comments on good features and areas for improvement:

6. Dictionary of Interesting Chemistry

Summary

Outline of the exercise

Students work in pairs to produce an entry for a new publication called *The Dictionary of Interesting Chemistry*. They are given a title, and are required to locate four references in order to produce an entry of about 300 words. Clues for the references are given, and will help them gain confidence in the use of BIDS and *Chemical Abstracts*. Although students are not required to analyse the papers in depth, they need to discern the key results and summarise them effectively in a high quality, illustrated dictionary entry. Some examples of work produced by students during trialling of the exercise are included in this book.

Key aims

■ to introduce the commonest methods of literature searching; and

■ to develop the ability to summarise information concisely.

Time requirements

■ Optional introduction to BIDS and *Chemical Abstracts* (approximately 1 hour)

■ Less than 1 hour (tutor contact time)

■ Approximately 20 hours private study

Timetable

Students might need an introduction to the use of *Chemical Abstracts* and BIDS. However, once they have received guidance on the use of the databases, only 5 minutes are required to explain the exercise; details can be found on the student handout. Students need about 20 hours to complete the exercise well, including 5 hours for finding the correct references, 7 hours for reading the papers and extracting the key information, and 8 hours for producing their dictionary entry in the correct format. Some students might need help from the tutor to locate their references. A reasonably imminent deadline is helpful – if students have other course commitments, a timescale of 2 weeks is appropriate.

S6 Dictionary of Interesting Chemistry

In this exercise you are required to produce an entry for the *Dictionary of Interesting Chemistry* that is to be published shortly. It will include some of the most interesting compounds, reactions, advanced materials, devices, and observations relating to chemistry. Examples include:

Compounds	anatoxin, cis-platin, nitric oxide, buckminsterfullerenes
Reactions	Wacker Process, photodissociation, polymerisation
Advanced Materials	liquid crystals, gels, dendrimers
Devices	microwaves, electrospray mass spectrometry
Observations	ozone hole, comets, the chemistry of odour

The entry should be about 300 words (with absolute limits of 200–400 words), including one graphic drawn by you (eg a chemical structure or scheme); one additional graphic (in any format) may also be added. It must fit on a side of A4, with typing in single-spaced 12 point Times font.

Each entry into the dictionary requires four references from primary journals, although a further review article or book reference may be added. Clues are given to guide you to three of the primary references, and a fourth reference must be chosen by you. The clues might include:

■ the first time the topic was published;

■ a key paper by one of the main researchers; and

■ one particularly interesting aspect of the topic, possibly a recent development.

The entry is to be pitched at the level of a chemist with an undergraduate degree – so you should be able to fully understand each other's entries in the dictionary. A typical entry might include:

Introduction (50 words)
eg Aspirin remains one of the most widely used painkillers. Despite having being developed about 100 years ago, new uses of the drug are still being discovered today, and interest in it has been renewed by its recent success in the treatment of stroke.
Discovery/Background (100 words)
Summary of the Chemistry (*eg* reactions, properties, applications) (150 words)
New aspect or unusual feature (100 words)

Sometimes scientific papers can be quite daunting, but you are not required to read them in detail. Just locate them and pick out the main observations, so that you (or others) can follow up on the details if necessary. A good textbook, appropriate dictionary (*eg* Dictionary of Organic Compounds), or good review article will provide a lot of the general information you need.

Topics and Clues

Topic 1

Microwave Synthesis

Your entry into the *Dictionary of Interesting Chemistry* must cover the application of microwave dielectric heating to the synthesis of solid-state materials, and its advantages over conventional methods. The report should include reference to the use of domestic ovens.

Reference one   *Nature* 1988, first author DRB.
Reference two   Same group, same year – an RSC journal.
Reference three *Chemical Communications*, 1996. Authors W and B.

Topic 2

$Rh_4(CO)_{12}$

Your entry into the *Dictionary of Interesting Chemistry* must describe this important binary carbonyl compound including the original synthesis, determination of its interesting structure and its catalytic applications.

Reference one   Booth *et al.*, 1968.
Reference two   Wei, the next year.
Reference three Japanese workers in 1991 using $Rh_4(CO)_{12}$
                 for a selective synthesis of what?

Topic 3

Anatoxin

Your entry into the *Dictionary of Interesting Chemistry* must cover the isolation, structure, pharmacology and one recent asymmetric synthesis of this toxin.

Reference one   What is the structure, and what does the acronym
                 VFDF stand for? Try Devlin 1977.
Reference two   How does it kill you? Try vol. 135 of a 1987
                 pharmacology journal.
Reference three What is the yield of a recent synthesis? Try Skrinjar, 1992.

Topic 4

Biological Applications of Poly(diacetylenes)

Your entry into the *Dictionary of Interesting Chemistry* must cover the detection of the influenza virus by functionalised PDAs. You need to describe the synthesis of PDAs and find the first use of them to detect the 'flu virus. Include the detection method used and illustrate their usefulness by describing further applications.

Reference one   American chemistry journal. 1993, authors include DHC.
Reference two   Same group, same journal, 1995.
Reference three More recently, the same group, but in a journal with a
                 medical leaning.

Topic 5

Main-chain Thermotropic LCPs

Your entry into the *Dictionary of Interesting Chemistry* must:

- define the key terms – LCP, thermotropic, main-chain;

- indicate the first synthesis of this class of compounds; and

- identify their useful properties and how they relate to applications.

Reference one   1975, authors AR and AS.
Reference two   1976, authors WJJ and HFK.
Reference three   Late 1980s by a single author, in a polymer
                       journal with an engineering slant.

Topic 6

Anisotropic Gels

Your entry into the *Dictionary of Interesting Chemistry* must describe the operating principle and primary application of these materials, which consist of low molecular weight liquid crystal (LC) containing a small proportion of networking agent. Also find a reference which describes their use with conventional LCs.

Reference one   1990, a physics journal, author RAMH.
Reference two   1992, same journal, 4 authors including Kim.
Reference three   Same journal, RAMH with another, early 1990s.

Topic 7

Photodissociation of Methane

Your entry into the *Dictionary of Interesting Chemistry* must describe the primary photodissociative channels in the photodissociation of methane in the vacuum UV region (100–300 nm). A number of possible reactions can occur which will produce CH_3 radicals, or CH_2 (methylene) with H or H_2. The key references show the development of the understanding of these processes.

Reference one   1962, authors' surnames M and M.
Reference two   1982, same journal, authors S and Black.
Reference three   Same journal, 8 authors including Morley, 1990s.

Topic 8

Cyclobutadiene

Your entry into the *Dictionary of Interesting Chemistry* must discuss the instability of free cyclobutadiene and explain how its co-ordination to metal complexes brings stability. Describe how complexes have been isolated and give the structure and reactivity of the co-ordinated ring.

Reference one   1956, in a journal of the Royal Society of Chemistry.
Reference two   1965, in the American equivalent.
Reference three   1975, author RP.

Topic 9

Vinyl Benzoate Polymerisation

Your entry into the *Dictionary of Interesting Chemistry* must discuss this anomalous reaction, briefly describing the type of polymerisation that is occurring.

Reference one   1959, Polymer journal – try a radical search!
Reference two   1949, it all starts from this.
Reference three 1959, a journal for big molecules. Authors V and S.

Topic 10

Suaveoline

Your entry into the *Dictionary of Interesting Chemistry* must cover the isolation, biosynthesis and the first total synthesis of this natural product.

Reference one   1972, journal name is 'TL'; suaveoline is an alkaloid.
Reference two   Phytochemistry journal, Enders (or Endress in BIDS) *et al* in the 1990s.
Reference three Mid 1990s, authors B and M.

Topic 11

Odour-Structure Relationships

Your entry into the *Dictionary of Interesting Chemistry* must comment on odour characteristics and how these relate to molecular structure. Describe the methods that are used to study this relationship (remember the American spelling).

Reference one   1967, a general scientific journal.
Reference two   1971, another general scientific journal, authors GFR and JIH.
Reference three 1992, pharmaceutical journal. Authors include LBK.

Topic 12

FR-900848

Your entry into the *Dictionary of Interesting Chemistry* must give the full structure and absolute stereochemistry of this antifungal antibiotic (it may be referred to as an antifungal agent), which were recently confirmed by elegant synthetic studies. Discuss the following: its source, biological activity and physicochemical properties.

Reference one   1990, authors are Yoshida *et al.*

Reference two   Barrett *et al* in 1995 describe approaches to the synthesis; AJPW is a co-author.

Reference three The following year, successful synthesis in American journal.

Topic 13

Cis-Platin

Your entry into the *Dictionary of Interesting Chemistry* must give this compound's structure and describe its primary application. Note its interaction with a key biological molecule and the development of related compounds for the same purpose.

Reference one   1969, in a general scientific journal.
                  Search for anti-cancer agents.
Reference two   mid-1980s, Lippard and others in another
                  general scientific journal.
Reference three   Early 1990s, same author with two others, reviews
                  this topic for inorganic chemists.

Topic 14

Polymer Dispersed Liquid Crystals

Your entry into the *Dictionary of Interesting Chemistry* must describe how the operation of displays incorporating these novel materials differs from those using conventional liquid crystals, and the advantages over LCDs. Explain how changes to the preparation methods and composition affect the physical form of the materials, which in turn affects the switching behaviour. Explain why they perform so poorly at certain frequencies of electric field. (NB: Early papers may refer to these materials by different names).

Reference one   1986, a physics journal, author Doane and three others.
Reference two   1994, chemistry journal, 3 authors including Amundson.
Reference three   Polymer journal, primary author's initials are ZZZ, 1992.

Topic 15

Danishefsky's Diene

Your entry into the *Dictionary of Interesting Chemistry* must describe the structure of this diene and its reactions with alkenes, carbonyls and imines (all acting as what?).

Reference one   An American journal. 1974, look for the man himself!
Reference two   It's him again – same journal, 1982 with JFK and SK.
Reference three   Same journal in the 1990s, but with new authors including
                  KI and Yamamoto

Topic 16

Reversible Oxygen Carriers

Your entry into the *Dictionary of Interesting Chemistry* must cover the principles behind the development of complexes which are capable of binding reversibly with O_2. Compare this to the naturally occurring complex and consider why the workers in this area adopted the structures of the various complexes selected.

Reference one   1973, American journal.
Reference two   Same journal – Busch and others in 1983.
Reference three   Large molecule journal in 1988 with 4 authors
                  including MO and HN.

Topic 17

Liquid Crystalline Metal Complexes

Your entry into the *Dictionary of Interesting Chemistry* must cover the structural types of metal complexes found in liquid crystalline (LC) systems and the reasons why metal complexes containing LCs (known as metallomesogens) are interesting.

Reference one   Nolte and others wrote this in 1988, in a Dutch journal.

Reference two   1991, two more transition elements (d^3 and d^8), Hoshino *et al.*

Reference three   1993, German journal – MJB *et al.* report which 'first'?

Topic 18

Electrospray Ionisation

Your entry into the *Dictionary of Interesting Chemistry* must cover the discovery of electrospray ionisation, including a brief summary of the principle. Also state the sensitivity of the technique and one recent application.

Reference one   Yamashita and colleague in 1984.

Reference two   1994, Emmett and RMC.

Reference three   1996 Clinical journal – author Sweetman.

Topic 19

Comets and the Origins of Life

Your entry into the *Dictionary of Interesting Chemistry* must describe how the impact of comets on the Earth's surface might have caused the basic building blocks of life to form. The papers relate to how comets colliding with Earth could have led to atmospheric chemistry from which organisms evolved.

Reference one   Mid 1980s paper, general science journal, BF plus 3 co-authors.

Reference two   Mid 1990s, a space journal, authors JO, TM and AZ.

Reference three   Late 1990s, another general science journal, authors CPM and WJB.

Topic 20

Ionic liquids

Your entry into the *Dictionary of Interesting Chemistry* must discuss the important qualities these liquids exhibit, the synthetic route to them, and their uses. You should also include recent developments concerning room temperature ionic liquids.

Reference one   Early 1980s, inorganic journal, first author JSW.

Reference two   A technology journal, 1990s, author KRS.

Reference three   A recent communication (late 1990s) by 5 authors.

Topic 21

Transition Metal σ-complexes

Your entry into the *Dictionary of Interesting Chemistry* must explain why transition metal σ-complexes are important models for the study of the activation (oxidative addition) of CH_4. Give a simple description of the bonding in such systems in molecular orbital terms and identify which heavier analogue of CH_4 has recently been characterised in a σ-complex.

Reference one	1984, American journal. Five authors including Swanson.
Reference two	Same journal, same year. Authors J-YS and RH.
Reference three	1995, same journal, but only two pages. Five authors, including Burns.

Topic 22

The Isomerisation of Carboranes

Your entry into the *Dictionary of Interesting Chemistry* must describe the mechanism of the isomerisation of icosahedral carboranes, which appears to be finally established; comment on the other major contributions over thirty years of speculation

Reference one	DG and JD publish important paper on carborane chemistry in 1963.
Reference two	Thirty years later, important calculations by Wales identify potential intermediates.
Reference three	1997, the first experimental proof of a mechanism is reported in a German journal by SD and others.

Topic 23

Nitric Oxide

Your entry into the *Dictionary of Interesting Chemistry* must answer the following questions concerning NO. In 1987, two groups, one led by Moncada, found that NO is an essential molecule in living organisms; who were the other groups, what does NO do, and what is its medical importance?

Reference one	1987, key publication by Moncada, with AGF and another as co-authors.
Reference two	DLHW and ARB review in 1993.
Reference three	1996 – Rossaint *et al.* describe NO in an enzyme-related journal.

Topic 24

Taxol

Your entry into the *Dictionary of Interesting Chemistry* must cover the source, biological activity, structure and physicochemical properties of taxol. Describe the mechanism of cytotoxicity, the source of the material and the dilemma attached to this. Find a total synthesis of this natural product and comment on the viability of the current source of supply.

Reference one	Authors Wani *et al* in the early 1970s.
Reference two	Review up to 1993 by Nicolaou and two colleagues.
Reference three	1994 was a first, reported by Holton *et al.* in two consecutive papers.

Topic 25

Organometallic Complexes of C_{60}

Your entry into the *Dictionary of Interesting Chemistry* must cover the development of the organometallic chemistry of C_{60}. Compare its behaviour to that of an electron deficient alkene or arene.

Reference one   Suggested phrases for searching: mono adduct, $[(C_6H_5)_3P]_2Pt$ and C_{60}; PJF *et al.*

Reference two $Ir(PPh_3)_2$ and C_{60}; ALB *et al.*

Reference three   A 1996 paper, comparing C_{60} to an arene in its bonding.

Topic 26

Symmetric Chlorine Dioxide

Your entry into the *Dictionary of Interesting Chemistry* must outline its preparation, detection in the stratosphere, and briefly describe its role in ozone depletion.

Reference one   RHD and colleague in 1953, try searching under chlorine(IV) oxide.

Reference two   1994, in a journal covering chemistry and physics, authors SH and JBN lead the way to its detection.

Reference three   1987, not a chemistry journal. Authors SS, GHM, RWS and ALS build on the previous work to find the answer.

Topic 27

Dendrimers

Your entry into the *Dictionary of Interesting Chemistry* must identify the nature of the structures, their unique aspects as large macromolecules of well defined molecular weight and the way in which these structures are synthesised.

Reference one   1990, a major paper with 170 references written by DAT and others.

Reference two   Same year, an alternative synthetic approach by Hawker and another.

Reference three   In the title of this slightly later paper, alternative names are included. Try arborols and cascade molecules; authors include HBM.

Topic 28

Ascaridole

Your entry into the *Dictionary of Interesting Chemistry* must cover the isolation, synthesis, pharmacological properties and rearrangement reactions of ascaridole.

Reference one   1971 paper by Bernhard *et al.*, in a biological journal.

Reference two   1953 paper by GOS, KGK and HJM.

Reference three   1979 paper brings more light to the chemistry of ascaridole. Two of the four authors have colourful names.

Topic 29

Artemisinin

Your entry into the *Dictionary of Interesting Chemistry* must cover its isolation, synthesis and pharmacological properties.

Reference one	1990s paper co-authored by Butler gives a general view.
Reference two	1980 paper with unspecified authors, determines the structure unambiguously.
Reference three	Elaboration of optically-active terpene precursor provides alternative route to title compound in early 1990s; 3 authors, including MAA in a chemical journal.

Topic 30

The Wacker Process

Your entry into the *Dictionary of Interesting Chemistry* must describe the conversion of ethene to ethanol and give details of the catalysts required (and their role in the reaction) and the optimum conditions.

Reference one	1959 paper with seven authors including Hafner, make the first progress using the Wacker catalysts.
Reference two	1962 paper gives more from the same team in the international edition of Reference one.
Reference three	1984, Japanese author adds the required details, in a review article. Search using the relevant metal.

T6 ■ Dictionary of Interesting Chemistry

This exercise is designed to be run soon after students have been introduced to literature searching for the first time. Input might come from library staff or an expert on information services, who often provide training sessions, videos and advice on the facilities available at a particular institution.

The exercise seems to work best if the students work in pairs, although it could easily be carried out individually. However, students gain confidence from working together on the information retrieval and they often produce higher quality submissions if they can critically discuss their entry. The allocation of suitable topics to pairs of students needs some preparation by the tutor.

Once the topics have been allocated tutor input is minimal, although additional clues can be given to students who are struggling (see notes below). The length of the report that is suggested in the student handout is deliberately short, as this exercise was designed to develop strategies for information retrieval. Moreover, the ability to produce short, clear summaries is a valuable skill.

Topics

Thirty topics are suggested, for which the following information is provided:

■ Handouts for students on each of the thirty topics;

■ A list which can be used to allocate titles to students (see next page);

■ Notes on where each of the clues leads (see Appendix B); and

■ The three references for each topic (see Appendix B).

By asking colleagues to contribute one (or two) new titles, and three appropriate references, additional topics can be generated.

Databases and journals
Although the exercise has been designed for *Chemical Abstracts* (in hard copy form) and BIDS (Bath Information Database Service), its framework can be adapted to introduce additional resources such as safety databases, CD-ROMs, other chemistry resources such as Beilstein and Gmelin, and the World Wide Web.

If students are expected to produce an extensive report, it will be necessary to check the availability of all of the referenced journals. Otherwise, the abstracts in BIDS or *Chemical Abstracts* are likely to contain the key points of the paper and will provide enough material for a shorter (approximately 300 word) report.

Student examples
Two examples of dictionary entries produced by students undertaking the exercise are included; whilst not perfect, they indicate the sort of entries that might reasonably be expected.

Titles of dictionary entries

Topic	Student(s)	Title of assigned topic
1		Microwave synthesis
2		$Rh_4(CO)_{12}$
3		Anatoxin
4		Biological applications of PDA
5		Main-chain thermotropic LC polymers
6		Anisotropic gels
7		Photodissociation of methane
8		Cyclobutadiene
9		Vinyl benzoate polymerisation
10		Suaveoline
11		Odour-structure relationships
12		FR-900848
13		Cisplatin
14		Polymer dispersed liquid crystals
15		Danishefsky's diene
16		Reversible O_2 carriers
17		LC-metal complexes
18		Electrospray ionisation
19		Comets and the origins of life
20		Ionic liquids
21		Transition metal sigma-complexes
22		The isomerisation of carboranes
23		Nitric oxide
24		Taxol
25		Organometallic complexes of C_{60}
26		Symmetric chlorine dioxide
27		Dendrimers
28		Ascaridole
29		Artemisinin
30		The Wacker process

Assessment

Tutor assessment is fairly straightforward and an appropriate assessment form is included in this pack. If the students can see each other's work their entries tend to be of a higher standard, especially if they are compiled into a 'Dictionary'. Peer-assessment can also be carried out, but it is important that a significant portion of the mark is set aside for location of the correct references.

Examples of student work

Taxol

has potent antileukemic and tumor inhibitory properties and is isolated from the stem bark of the Pacific yew, *Taxus brevifolia*. First isolated in 1969, the cytotoxic behaviour was first reported by Wani and Hall[1] with the first total synthesis reported independently by Nicolaou[2] and Holton[3]. It has recently completed its clinical trials and is used in the treatment of breast and ovarian cancer. [4]

All plant and animal cells that have a nucleus, contain a protein called tubulin, the purpose of which is to form microtubules, which act as cell templates and push apart dividing cells. Most ordinary body cells divide only very infrequently, whereas cancer cells divide very rapidly. Taxol unlike other anti-cancer agents actually stimulates the formation of microtubules but still inhibits cell division.

1. Taxol, R^1 = Ph, R^2 = OAc   2. Taxotere, R^1 = ^tBuO, R^2 = OH

The Pacific yew tree from which taxol is derived is slow growing, small in height and therefore approximately 3000 trees are required per kilogram, enough to treat only 500 patients. As a result alternative sources have been sought in the form of Taxotere (essentially taxol but with a different side chain protecting group at nitrogen), currently produced synthetically by adding a side chain to 10-deacetyl baccatin III, which is readily available from the renewable needles of the European yew, *T. bacatta*.

1. M.C. Wani *et al.*, *J. Am. Chem. Soc.*, 1971, **93**, 2325.

2. K.C. Nicolaou *et al.*, *Angew. Chem. Int. Ed. Engl.*, 1994, **33**, 15.

3. R.A. Holton *et al.*, *J. Am. Chem. Soc.*, 1994, **116**, 1597.

4. P. Jenkins, *Chem. Br.*, 1996, **32**, 43.

Dawn Robinson and Trish Drennan

Microwave Synthesis

It has been known for a long time that we can use electromagnetic waves to bring molecules into excited rotational states, this aspect of physical chemistry being utilised domestically in the microwave oven to heat food. However,

some years ago, chemistry took the discovery and put it back into the hands of the scientist to heat their reactions. Since this first step many reactions have been conducted using microwave ovens, domestic and specially designed, with very intriguing results.

We know that water molecules are the target of the domestic oven but any molecule with a permanent dipole is susceptible. The average time taken to rotate a molecule through one radian (correlation time) is similar enough in most molecules to be excited by the frequency of the domestic microwave 2.45 GHz. The microwave energy is efficiently changed into heat and superheating is easily achieved at quite low pressures. This can give amazing improvements in rates of reactions of organic species eg the Diels-Alder reaction of maelic anhydride with anthracene in diglyme (bp 162 °C) takes one minute in the microwave and gives 90% product whereas it would take ~90 minutes conventionally.

In the solid state, coupling to microwaves is not usually observed but conducting and semiconducting solids can couple through the movement of electrons or ions. As the temperature increases the efficiency of coupling increases, so temperature increases further and so on until a thermal runaway is observed. This leads to melt but when the microwaves are switched off the solid cools very quickly. This technique can be used to create high temperature ceramic superconducting materials.

Many of the inorganic oxides from which the ceramics are made do not absorb microwaves, but some of them do (see table 1), meaning a reaction mixture of some absorbent types mixed with transparent ones will still heat very rapidly. Where conventional techniques could take from 12–24 hours, the microwave method of melt and quench can produce pure samples in a matter of minutes.

Table 1	
MW absorbing	**MW Transparent**
ZnO	CaO
V_2O_5	TiO_2
CuO	Al_2O_3
MnO_2	Fe_2O_3

By the amazing claims suggested in chemical literature, it is obvious that this area of chemistry could prove to be of significant importance in the future of chemical synthesis.

1. D.R. Baghurst and D.M.P. Mingos, *J. Chem. Soc., Chem. Commun.*, 1988, 829.

2. D.R. Baghurst, A.M. Chippendale and D.M.P. Mingos, *Nature*, 1988, **332**, 311.

3. C. Wu and T. Bein, *J. Chem. Soc., Chem. Commun.*, 1996, 925.

4. D.M.P. Mingos, *Advanced Materials*, 1993, **5**, 857.

Raymond Adamson and Atif Shafiq

RS•C

COMMUNICATING CHEMISTRY

Cis-Platin

Felix Onasanwo and Jonathan Hussein

Cis-Platin is a chemically important anti cancer drug especially effective for the management of testicular, ovarian, head and neck tumours. Considerable evidence points towards DNA as being the main target of cis-platin in the tumour cell. Therefore attention has been placed on the difference between the adducts formed with DNA between cis- and trans- platin.

It was discovered that certain platinum containing compounds completely but reversibly inhibit cell division in the gram negative rods. Tests for anti tumour activity proved that these platinum containing compounds inhibit sarcoma 180 and and leukaemia in mice. The results showed that the compounds were effective in inhibiting the tumours and the mice remained free from tumours for up to six months living normal healthy lives, however at this stage no knowledge of the fate of the injected compounds or the mechanism of action against the tumour cells was known.

Cis-platin reaction with DNA involves the loss of two chloride ions and the formation of two Pt-N bonds to the N(7) atom of two adjacent guanosine nucleosides on the same strand. For stereochemical reasons this cannot take place with trans-platin as the formation of 1-2 inter strand crosslinks cannot take place hence the trans- isomer of platin is inactive in this field. Co-ordination of the platinum to DNA takes place in two stages, firstly the formation of the mono functional adducts primarly to the N(7) position of guanine or adenine. These react further to form bifunctional adducts at N(7) positions of nearby guanines and to a lesser extent adenines. If the co-ordinated nucleotide bases are on the same strand of DNA an inter strand crosslink is formed, however if the platinum links to two bases of opposite DNA strands the result is an inter strand adduct. The adduct formed with cis-latin is:

cis[Pt(NH$_3$)$_2$ [d(GpG)]

Crystals of the adduct formed have been obtained and used in an X ray crystallographic study to elucidate the molecular structure to atomic resolution.

Since discovery of the antineoplastic activity of cis-platin considerable progress has been made in understanding how platinum complexes bind to DNA. Further knowledge of how anti-tumour drugs work would have major implications for further design and improvement of these drugs, it is also becoming possible to design complexes so as to bind to DNA in a predictable way paving the way for the development of new drugs for the hopeful cure of all types of cancer.

References
1. *Nature* 1969, vol **222**, pg 385.

2. *Science* 1985, vol **230**, pg 412.

3. *Progress in Inorganic Chemistry* 1990, vol **39**, pg 477.

4. *Co-ordination Chemistry Reviews* 1990, vol **100** pg 293.

Comets and the origins of life

Neil Polwart and Martin Melia

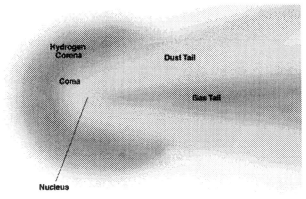

Composition of Comet 1

A comet is composed of a ball of frozen gases. The forces of a comet colliding with the Earth's surface are thought to be strong enough to cause chemical reactions that may have created the building blocks of life.

All living creatures, DNA, enzymes and proteins are built up from molecules containing the $>NH_2$ group, such groups originate from ammonia – but how the ammonia was first created has puzzled scientists for years. Scientists including a team working at NASA, have bombarded a mixture of gases (at low temp) with high energy lasers designed to simulate the forces experienced by the small molecules as they hit the Earth's surface.

Initially shocking of the gases which are rich in methane produces hydrogen cyanide and acetylene but further shocking with the lasers produces ammonia. It is possible that this ammonia eventually ended up as amino acids and became the building blocks of life. Theories along these lines have been suggested since 1961 but recent work suggests that this would not occur if there was high CO_2 concentration in the atmosphere -- as had previously been suggested. Perhaps life could exist on other planets – if the collisions with comets were repeated elsewhere.

$$H-C\equiv N \qquad H-C\equiv C-H$$

Some interesting references:

1. Organic synthesis in Experimental Impact Shocks; *Science*, 1997, **276**, 5311, p. 390. C.P. McKay & W.J. Borucki.

2. Chemical Effects of Large Impacts on the Earth's Primitive Atmosphere; *Nature*, 1986, **319**, p. 305. B. Fegley (Jr), R.G. Prinn, H. Hartman, G.H. Watkins.

3. Probing the presently tenuous link between comets and the origins of life; *Origins of Life*, 1982, **12**, 2, p. 125. R.W. Hobbs, J.M. Hollis.

4. Comets and Life in the Universe; Advances in *Space Research*, 1994, **15**, 3, p. 81. J. Oro, T. Mills, A Zazcano.

5. Comet Impacts and Chemical Evolution of the Bombarded Earth; *Origins of Life and Evolutions of the Biosphere*, 1992, **21**, 5/6, p. 317. V.R. Oberbeck, H. Aggarwal.

Or check out the NASA home page at or the Science home page at: http://www.nasa.gov http://www.science.com (password needed; 18/04/97 edition).

Tutor's guide to the clues

The clues to the references should guide the students to a small number of possible references, from which the one required should be fairly obvious. Pre-1981 references must be found from *Chemical Abstracts*. References from 1981 onwards are usually most easily found using BIDS, and descriptions that successfully locate each reference are given in the Table for each topic in Appendix B.

Assessment form

Title of dictionary entry:

_____

Authors:

_____

Assessment criteria:

■ Did the report start with a clear summary of the background? /10

■ Was the report well written (grammar and style)? /10

■ Was the science well explained? /10

■ Was the presentation good (including graphics)? /10

■ Was it the right length, and pitched at the right level? /10

■ Were the key references correctly identified (tutor only)? /30

Total mark _____/80

Total mark as a percentage _____%

What was your overall impression of the quality of this report as % mark?

(Excellent >80%; very good 70-80%; good 60-70%; average 50-60%; poor 40-50%; very poor <40%)._____%

If appropriate, amend the marks given until the 'impression' mark awarded matches the total percentage mark.

Final mark _____%

Brief comments on good features and areas for improvement (especially where marks awarded to each of the above are either very high or very low):

7. Hwuche-Hwuche Bark

Summary

Outline of the exercise

Students are presented with a specific chemical problem set in an industrial context. They work in teams representing subsidiaries of a large company called ACE. In the scenario used here an academic chemist claims to have identified a tree bark with remarkable plant growth-enhancing properties. The academic has approached the 'companies' and they must therefore work against a series of deadlines and in competition with each other in order to:

■ secure a deal with the academic;

■ identify six compounds extracted from the bark using spectroscopic data;

■ determine the best way to synthesise the active component; and

■ explain the success of their new product in an oral presentation.

This pack contains a complete set of student handouts for the above scenario, examples of six compounds and their spectra, interpretations of these spectra and detailed tutor's guide for running the exercise. Ideas for further scenarios are also given.

Key aims
■ To develop the skills of team working and problem solving.

■ Time management, information retrieval, and presentation skills also feature strongly in the exercise.

Time requirements
■ Approximately 2 hours tutor contact time (see table below)

■ 6 hours private study

■ 8 hours total student time

Timetable

This exercise requires 8 hours of student work in its current form. It has been run as a one-day event, during the course of a week (using 4 lecture slots), and (in an extended form) over a university term/semester. Because the teams undertake extensive private study, the exercise can be adapted to fit the time constraints of any institution. A basic timetable is given below:

Session	Time	What the tutor does	Handout	What the companies do next
1	30 mins	Sets the scene	1	Prepare bids and submit them (1 hour)
2	5 mins	Gives out spectra	2	Solve structures and submit them (3 hours)*
3	30 mins	Announces 'news'	3a–d, 4	Prepare presentations (2 hours)
4	1 hour			Give presentations

(* An additional 'help' session can be added – see tutor's guide)

S7 Hwuche-Hwuche Bark

Student Handout 1

The product

Dr Barley is a lecturer in the Agricultural Science Department at the University of Rutland. During a trip to Eastern Asia last year, he came across a small community that spread the shredded bark of a local tree – the Hwuche-Hwuche tree – on the soil where they plant their crops; there appeared to be a significant increase in the speed with which some of the crops germinated and grew in this area, but the locals had managed to keep the nature of their 'fertiliser' a secret.

One of the crops which appears to respond to the application of the bark is closely related to maize. On his return to England with samples of the bark, Dr Barley carried out experiments to ascertain whether the development of maize plants could be accelerated using the bark – the results were extremely encouraging, and he is hoping that the material could have commercial value.

Unfortunately, the tree is not widespread, and it would not be viable to use the bark itself, even if further studies confirm its effectiveness. However, Dr Barley would be willing to give samples of the bark (*ca* 1 kg) to your firm for further studies, providing that a suitable agreement is in place – although he will be approaching other companies if you do not respond positively enough. Whether or not he makes an agreement with your firm will depend on long- and short-term financial incentives you offer, as well as on the quality of the associated research programme.

The company

Your company is a subsidiary of a much larger European agrochemical company called ACE. Nevertheless, you are allowed to operate fairly independently, and you have specific products that are entirely under your control. You also have a skilled R&D department, which could easily undertake a medium-term research programme without recourse to ACE. However, if you wished to undertake the large scale production of a new product you would need initial funding from your parent company – or, if the market were large enough, ACE might wish to make the product at one of their larger European plants. Your company has the following profile:

Products	Agrochemicals, mainly bulk fertilisers
Employees	Management and Marketing (8)
(total number: 80)	Production (40)
	Packaging and Distribution (15)
	Maintenance and Services (9)
	R&D (8)
Turnover	ca £25,000,000 p.a.

Initial assignment

You are all members of the research committee, set up by the Managing Director, which meets every month to plan the use of resources. The research plans are influenced by many factors, so the committee members (who are all trained chemists) are the Managing Director (Chairperson), Finance Director, Sales Director, Research Director (overseeing all R&D), and Chief Chemist(s). In response to Dr Barley, you must discuss:

1)   What your immediate short-term experiments should be;

2)   A likely deal for Dr Barley; and

3)   What longer-term plans should be considered if the Hwuche-Hwuche bark looks promising.

Submit the names of the research committee of your company, a deal for Dr Barley, and an outline of your longer-term plans by the deadline stated, in a sealed envelope, with your company's name on the outside.

Student Handout 2

Your proposal to Dr Barley has been accepted. He may have accepted offers from other companies, so there is a possibility that you will be working in competition with them. You have secured about 500 g of the Hwuche-Hwuche bark.

To save time, your top chemists have tried to isolate the active component from Hwuche-Hwuche bark, whilst the biologists have been checking that the whole bark does accelerate plant growth. After one month of extraction and chromatography, you believe that you have crude material containing the active component, some of which is sent for biological testing. The rest is found to contain several compounds, and a further month of careful chromatography has allowed you to isolate about 10 mg of each of the six major compounds in (almost) pure form. This is enough for full analytical data to be collected.

Your urgent priority now is to assign the structures correctly – accuracy is more important than speed, but there is nevertheless pressure on you to solve the structures quickly.

■ Remember that you are a team – you may wish to divide up the structure solving and then get together to check the proposed structures.

■ Consultants may be available to offer advice on the data.*

■ You must submit your six proposed structures in a sealed envelope by the deadline given.

*optional "help" workshop, at which specialists offer advice are availabe.

Spectra for the unknowns are given in Appendix C

Student Handout 3a

News from the plant biologists

The active component in the bark is compound **B**. It is three times as effective (weight for weight) as the market leader 'megasprout', for which the active component costs ca £50 per kilo. As with many fertilisers/growth promoters, the active component in 'megasprout' is highly potent, and is diluted down many thousand-fold in the actual product that is spread on crops.

Literature information on B

The commercial viability of the new fertiliser must be established, including details of its synthesis, and the cost of its production.

A quick literature search on compound **B** has provided some useful information. Using this information two possible synthetic routes have been outlined on the following handouts. You must check whether a better synthesis has been published, and also whether there are patents and biological properties reported for compound **B**. In order to cost it, you can assume that bulk materials could be obtained at a third of the Aldrich catalogue price. Gases are very cheap, but their containers are expensive, so bulk prices drop hugely – ammonia costs about £1 per kilo, and hydrogen about £8 per kilo in bulk.

Synthesis of B, Route 1

Ph——CH$_2$CN

Step 1 | 1. Na/NH$_3$ (l)
2. ClCH$_2$CO$_2$Na
3. HCl (aq)

Ph——CH$_2$CN
|
CH$_2$CO$_2$H

Step 2 | H$_2$/Raney-Ni/
4% NH$_3$ (aq)

Ph——CH$_2$CN
|
CH$_2$CO$_2$⁻.NH$_4$⁺

Step 3 | Distil *in vacuo*
at 210–220 °C

Ph

(pyrrolidinone structure)

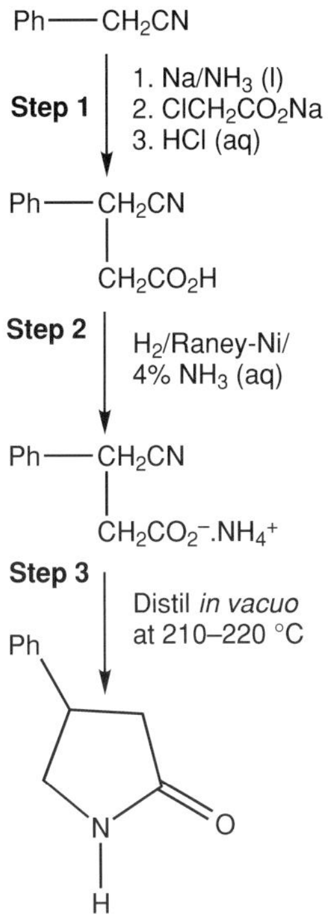

Step 1
This conversion was reported to proceed in ovarall yield of 75% (see *Chemical Abstract* below).

Step 2
Isolated yield and full experimental details were not given for this step. Similar work in 1976 gave the following results (G. Bettoni *et al.*, *J. Heterocyl. Chem.*, 1976, **13**, 1053):

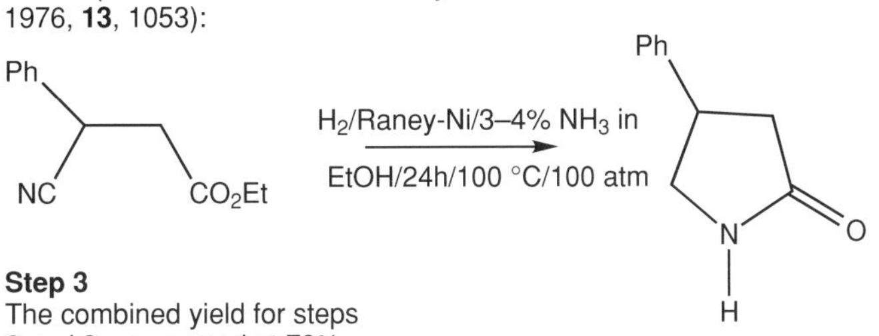

(structure: Ph / NC / CO$_2$Et)

H$_2$/Raney-Ni/3–4% NH$_3$ in
EtOH/24h/100 °C/100 atm →

(Ph-substituted pyrrolidinone structure)

Step 3
The combined yield for steps 2 and 3 was quoted at 79%

Published data relating to synthesis of B, Route 1 from *Chem Abs*. 1967, 66, 94779y

94779y Synthesis and reactions of β-phenyl-β-cyanopropionic acid. A. G. Chigarev and D. V. Ioffe (Inst. Toksikol., Leningrad). *Zh. Org. Khim.* **3**(1), 85–8(1967)(Russ). Dry ClCH$_2$CO$_2$Na (116 g.), m. 215°, was added to (PhCHCN)⁻Na⁺ (prepd. in situ from 117 g. PhCH$_2$CN and 25.3 g. Na in 1.5 l. liquid NH$_3$) and stirred 1 hr. to give 74–6% PhCH(CH$_2$CO$_2$H)CN (**I**), m. 74–6°. Crude **I** boiled 4 hrs. with concd. HCl gave 90% PhCH(CH$_2$CO$_2$H)CO$_2$H (**III**), m. 165–7°. A soln. of 141 g. **III** in 110 ml. 26% aq. NH$_3$ was evapd. to dryness, the residue melted and kept 30 min. at 200–20°, cooled, dissolved in hot EtOAc, filtered, and crystd. to give 75% phenylsuccinimide (**V**), m. 89–90°, which could be also obtained by heating PhCH(CH$_2$-

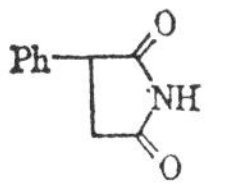

(structure: Ph / NH (V) / O, O)

CO$_2$H)CONH$_2$ [obtained from **I** by the Anschutz procedure (*CA* **1**, 2702)]. Redn. of **V** with LiAlH$_4$ gave a 66.5% yield of 3-phenyl-pyrrolidine, b$_2$ 92–4°, n$_D^{20}$ 1.5545 (picrate m. 159–61°). Esterification of **I** with MeOH with a catalytic amt. of H$_2$SO$_4$ gave 69% PhCH(CH$_2$CO$_2$Me)CN, m. 50–2°. Hydrogenation of **I** in 4% aq. NH$_3$ with Raney Ni gave PhCH(CH$_2$CO$_2$NH$_4$)CH$_2$NH$_2$, which on heating at 210° cyclized to β-phenylbutyric acid γ-lactam, m. 75°, b$_2$ 178–82°, in 79% yield calcd. on **I**.   CPJR

Student Handout 3c

Synthesis of B, Route 2

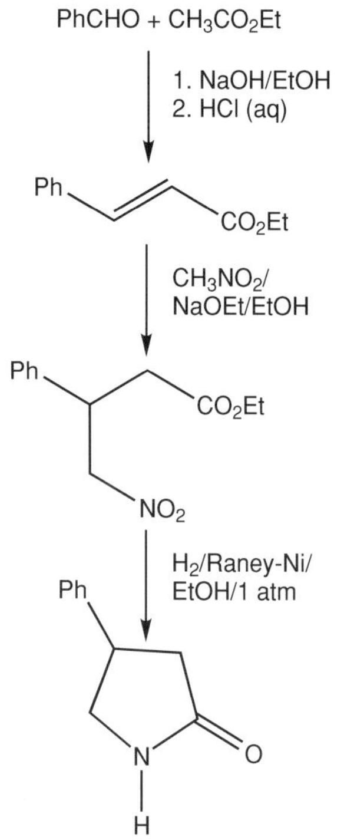

PhCHO + CH₃CO₂Et

Step 1
This conversion was reported
to proceed in ovarall yield
of 75% (see procedure in Handout 3d).

1. NaOH/EtOH
2. HCl (aq)

Ph $\diagup$ CO₂Et

CH₃NO₂/
NaOEt/EtOH

Step 2
This should proceed by conjugate (Michael)
addition of the CH₂NO₂ anion, and should require
1 mol equivalent of base. This type of reaction is
well known, and yields of around 80% are expected.

Ph $\diagup$ CO₂Et

NO₂

H₂/Raney-Ni/
EtOH/1 atm

Step 3
This step is in the literature and is claimed
to proceed in 75% yield (J. Cologne *et al*.,
Bull. Soc. Chim. Fr., 1962, 598).

Ph

N—O
|
H

Student Handout 3d

Published data relating to synthesis of B, route 2

From: *Organic Synthesis,* Coll. Vol. I, 1932, 252.

ETHYL CINNAMATE

$$C_6H_5CHO + CH_3CO_2C_2H_5 \xrightarrow{\quad C_2H_5ONa \quad} C_6H_5CH{=}CHCO_2C_2H_5 + H_2O$$

Submitted by C. S. Marvel and W. B. King.
Checked by Henry Gilman, R. E. Fothergill and R. E. Brown.

1. Procedure

In a 2-l. two-necked flask fitted with a short reflux condenser and mechanical stirrer (Note 1) are placed 400 cc. of dry xylene (Note 2) and 29 g. (1.26 atoms) of clean sodium (Note 3) cut in small pieces. The flask is surrounded by an oil bath and heated until the sodium has melted. At this point the stirrer is started and the sodium is broken up into very small particles (Note 4). The oil bath is removed, but stirring is continued until the sodium has solidified in very fine particles. The xylene is then poured off, and to the sodium is added 460 cc. (4.7 moles) of absolute ethyl acetate (Note 5) containing 3–4 cc. of absolute ethyl alcohol (Note 6). The flask is quickly cooled to 0° and 106 g. (1 mole) of pure benzaldehyde (Note 7) is added slowly from a separatory funnel while the mixture is stirred. The temperature is held between 0 and 5° (Note 8). The reaction starts as soon as the benzaldehyde is added, as is shown by the production of a reddish substance on the particles of the sodium. About one and one-half to two hours are required for this addition. The stirring is continued until practically all of the sodium has reacted (one hour after all the aldehyde has been added).

When most of the sodium (Note 9) has disappeared, 90–95 cc. of glacial acetic acid is added and the mixture is carefully diluted with water. The ester layer is separated, the water layer is extracted with about 25–50 cc. of ethyl acetate, and the combined ester portions are washed with 300 cc. of 6 N hydrochloric acid and then dried with sodium sulfate. The ethyl acetate is distilled from a water bath and the remaining liquid is transferred to a Claisen flask and distilled from an oil bath under reduced pressure. A small fraction comes over below 128°/6 mm. and is discarded. The ethyl cinnamate (Note 10) boils at 128–133°/6 mm.; 168–173°/46 mm. The yield is 120–130 g. (68–74 percent of the theoretical amount). (NB The "notes" are not reproduced here.)

Student Handout 4

Urgent message from ACE

Your parent company ACE has been informed of the potential of Hwuche-Hwuche bark, but is currently reviewing all its major R&D commitments. They plan to announce a series of investments and cuts in about a month's time and they have asked your management team to give a presentation to the ACE board, outlining the potential of the Hwuche-Hwuche bark project. This presentation must last about five minutes (not less than four or more than seven minutes), and will be given by one or two of your management team. The presentations will be in the next few days – check when you are timetabled to speak. Materials for preparing overheads are available.

Presentations might concentrate on two or three of the following points:

■ How successful you were at isolating and identifying **B**;

■ A plan for the next phases of its development and any investment requirements;

■ How you plan to make it, and how much it will cost (including scale up savings);

■ How it compares with rival products (for example, cost and properties) or how you might market it – advertising ideas might come in here;

■ How much profit it might make (and whether Dr Barley will get a big slice of it).

Hwuche Hwuche bark

This exercise requires three plenary sessions of between three and thirty minutes, and a longer session for final presentations. It can be run over a week, as a one-day exercise, or spread over several weeks, although in this case it may lose some impact. Detailed advice on each session is given below. Although the exercise can be run by a single tutor, there are two instances where additional support might be helpful – during session 1, if the role play is to have maximum impact, and during the optional help session when the presence of three or four 'consultants' is beneficial. Six 'unknown' compounds can be found in this pack; five of the six (excepting the active compound) are available from Aldrich, and spectra and their interpretations are also included (see Appendices C and D).

Session 1

5 mins	Exercise introduced
5 mins	Students split into (pre-arranged) groups
15 mins	Handout 1 provided, and companies encouraged to start the exercise
10 mins	Plenary feedback session
	Deadline for bids announced

The tutor's introduction to this exercise can be used to set the problem in context and all of the information necessary is found on the student handout (Handout 1). During trialling it was found to be particularly effective if Dr Barley was introduced in a role-play, in which he enters the "office" of the managing director of the company; these roles can be played by the tutor and a colleague respectively. 'Companies' should have approximately six students each, and should assign themselves a company name and identify roles within the group (as described on the handout).

The plenary session can be used to gather feedback from the students concerning the short-, medium- and long-term considerations of the project, and all groups should be called upon to put forward suggestions. These might include, for example, getting the deal with Dr Barley, details of plant testing, identifying the components, lead compounds, development of a market compound, factors influencing success or otherwise (including cost, toxicology, and environmental issues) and time-scales. A summary of the points made in the discussion, given at the end of the plenary session, can be useful.

Finally, announce a deadline for submitting bids.

Session 2

This session can be very short, and simply requires that Handout 2 and spectra are given to the students. Handout 2 tells each of the companies that their offer to Dr Barley has been accepted, and introduces the problem solving aspect of the exercise (**spectra for this exercise can be found in appendix C**). An element of urgency, and therefore pressure, can be brought in if the companies are made to realise that they are in competition with each other. The emphasis should be on the importance of finding the correct structures (for which marks are given) and on team work.

Optional workshop with consultants

An extra session offering guidance on identification of the unknown compounds can be useful. Students should not be told whether their structures are right or wrong. Other colleagues, postgraduate students or research assistants might be involved as consultants. Interpretations of the spectra for the six unknowns in this pack are provided in Appendix D.

Hwuche Hwuche Bark – structures of substances isolated from bark

A   B   C

D   E   F

Session 3

5 mins	Identify unknown structures and request IUPAC names for them. At least one copy of Aldrich should be available per company
15 to 25 mins	Collect submissions of compound names Distribute handouts 3a–d Distribute handout 4 while companies are studying handouts 3a–d

The purpose of this session is initially to name the compounds and find out which of them are commercially available. The correct structures should be displayed so that all of the students are working on the same compounds at this stage. Of the six compounds used in this teaching pack five can be located in Aldrich, and the name for the sixth (the active compound) should be determined. Marks should be awarded for correctly naming each of the structures.

Handouts 3a–d explain the second part of this session. Students are now required to examine the literature provided on the synthesis of the active compound and decide how it might best and most cost effectively be prepared commercially. They are then required to give a presentation (see Handout 4) which addresses the industrial production of the fertiliser.

Session 4

Students give their presentations during this final session, the length of which depends on the number of groups involved. Presenters should remain outside the room until they have given their presentations, as it may be unfair for a later presenter to see an earlier presentation. The managing director of ACE can chair the session, and provide a short summary of the exercise at the end.

An assessment scheme is provided in this pack, based on various aspects of the exercise, and prize(s) can be awarded to the best companies; if students have taken part with enthusiasm, this helps the exercise to end on a high.

Adapting the exercise

This exercise can be easily modified. The basic format breaks down into:

a) The introduction, in which a commercial possibility (or problem) is set in context, and companies are established. Some general discussion helps the companies to get started, to identify roles, and to think about some longer-term issues.

b) An urgent problem solving session tackling specific chemical questions – identifying unknown compounds (of appropriate difficulty), a literature search, or the analysis of data relating to a process.

c) A solution to the problem can be identified within the overall context of the exercise.

d) A final presentation – a report, an oral presentation, or a poster.

Additional scenarios

Outlines are given below for various scenarios which could be used in a similar manner to the Hwuche-Hwuche bark exercise developed here.

■ Contamination scare
A company employee has spotted several dead fish near the (harmless?) effluent outflow. The company must respond before it becomes public. What should they do?

Session 1
Students are divided into two groups representing companies and environmental control units respectively.

A range of possible chemicals from the fictitious company, and their toxicity levels, should be provided by the tutor. Students need to identify suitable analytical methods which could be used to identify these. Methods chosen should be practicable for the undergraduate laboratory. Contaminants chosen could be organic or inorganic, highly toxic trace materials requiring tricky physical chemistry to quantify, or less toxic compounds present in larger amounts.

Session 2
In the laboratory, samples are provided which contain acceptable (measurable) levels of three contaminants, but dangerous levels of one. Students must identify this contaminant. Data could be provided if an appropriate laboratory session were not available.

Session 3
Companies must identify how to reduce the level of the contaminant in the effluent by looking at the industrial process (provided by the tutor). Environmental units must prepare a one-page report outlining their case for prosecuting the company.

Session 4

The problem has gone public. In the final session, a series of short TV interviews take place between an interviewer (environmental unit representative) and a company representative – the former asking questions to show how awful the company is, and the latter trying to defend their position and subsequent action.

■ Cheaper production costs

This exercise could be based on organic, inorganic or physical chemistry, and is a much simpler exercise to run. It involves identifying a synthetic process [for example, an actual commercial procedure (perhaps an old one), or a suitable synthesis that is not actually commercial], which could be improved by:

a. changing the synthetic route (new organic synthesis);

b. using a better inorganic reagent or catalyst; or

c. modifying the reaction conditions.

Session 1

The exercise starts with an outline of the process, and a discussion concerning the process. From the information given, the teams must estimate the cost of making the product.

Session 2

An emergency memo arrives indicating that a rival is undercutting their price, and they must reduce their cost by 30%. The teams need to use literature, catalogues, and their skill to develop a modified procedure that reduces the cost as much as possible.

Session 3

It is useful to have a halfway workshop in which groups can discuss their ideas with a consultant (ie with a tutor). It is then revealed that company cuts are due, and only the really competitive products will survive.

Session 4

At a series of presentations, the company management (*ie* the tutor) will hear proposals from teams about making their product more competitive. Assessment can be based on presentations and/or a single A4 report, and/or a poster. The winning team is the one which presents the most convincing case that they can reduce the cost.

■ Silent witness

Teams of students represent chemists in a forensic science department. There are any number of possibilities for inventing a crime, and getting the teams to try and solve it using chemical evidence, provided as the exercise unfolds. It would be possible to end such an exercise with a court case, particularly if there was conflicting evidence that defence and prosecution teams could use.

■ New pharmaceutical leads

This exercise could closely follow the Hwuche-Hwuche bark format presented in this chapter, with medicinal chemistry considerations replacing agrochemical factors.

Extending the exercise

Some possibilities for extending the exercise are:

■ Addition of more (and/or harder) unknowns;

■ Addition of a financial management component (*eg* companies have a budget from which they must buy spectra or consultancy help);

■ A substantial literature search, to explore syntheses and properties of an active compound;

■ Production of a final report (to a specific format).

Subsequent exercises

The 'Hwuche-Hwuche Bark' project establishes 'companies' that work well as teams (even though there may be some clashes of personality), and these teams can be used in the three subsequent exercises, extending the basic scenario:

■ Annual Review Presentations – where students give individual talks to their colleagues, as part of the company's annual review of its graduate employees.

■ Interviews – where the companies advertise, interview, and appoint a new employee; all of the students must apply for a job in another company.

■ Posters – where the company teams each prepare a poster on a new potential area of research.

These are stand-alone exercises, but the use of the companies from the 'Hwuche-Hwuche Bark' exercise adds realism, encourages development of team skills, and saves time.

Assessment

An assessment scheme is provided below.

a) Offer to Dr Barley (/10)
Top marks should be awarded for offers that do not cost too much, but would be likely to secure the material.

b) Structures (/24)
For each structure: 4 marks (correct), 2 marks (close), or 0 marks (incorrect).

c) Names of the compounds (/6)
For each compound correctly named – 1 mark.

d) Presentations (/20): (i) Quality of science (/8)
(ii) Quality of presentation (including visual aids) (/8)
(iii) Did they engender confidence and enthusiasm? (/4)

Company	Offer	Structures	Names	Presentation	Total (/60)

RS•C

8. Annual Review Presentation

Summary

Outline of the exercise
The aim of this exercise is to develop students' oral communication skills. The scenario used is that of taking part in a company's annual review of its graduate employees for which students are required to give a short presentation to their colleagues. They must choose a topic from current literature relating to a novel area of chemistry with commercial potential and talk for approximately six minutes. In order to increase students' awareness of what constitutes good and poor presentations, they are also required to consider their colleagues' talks and provide them with feedback.

Key aims
- to develop oral presentation skills, including creation of effective visual aids;

- to understand and explain a new piece of chemistry; and

- to carry out and benefit from peer assessment.

Time requirements
- 2.5 hours tutor contact time

- 9.5 hours private study

- 12 hours total student time

Timetable
The following timetable is suggested:

30 mins	Introduction (lecture slot)
9-10 hours	Students prepare their talks (private study, over a period of approximately 1 week)
2 hours	The talks are given in 1 hour workshops; students attend two of these – one as a presenter and one as an assessor

S8 Annual Review Presentation

The annual review procedure that your company carries out of all its employees is fast approaching, and you are hoping for promotion (or at least a salary rise). As part of their policy to identify potential group leaders, and to encourage good communication within sections, all graduate employees are being asked to give a short talk (approximately six minutes in length) on an exciting new piece of chemistry that might be developed by your company to commercial advantage. The Royal Society of Chemistry's magazine, *Chemistry in Britain*, newspaper science pages, and Internet pages should provide potential areas of interest. The presentation will be given to a review panel of chemists who know nothing about the details of this area. Your talk must achieve the following three objectives:

■ make your audience interested and enthusiastic in your topic;

■ explain the science clearly and accurately; and

■ demonstrate the commercial potential.

You must attend two seminars – one as a speaker, and one as a member of the audience providing feedback. Once you have chosen a topic to talk about, write an informative (but catchy) title on the notice-board – make sure that nobody in either of the two seminar groups you will be attending has chosen the same topic. Seminar times will be posted on the notice-board. You may use up to three (but no more) overheads during your presentation. Freehand drawings are fine and are often clearer and more colourful than computer print-outs.

Begin to plan your talk and visual aids as soon as possible – it's often helpful to discuss ideas with colleagues. Most good presentations have a clear introduction, followed by the main content of the talk and a brief conclusion. It is important to keep to time, so check the length carefully. Try not to say too much, and ask a friend to listen to it in advance. Look at the presentation styles of your lecturers (both good and bad!) to pick up hints.

T8 Annual Review Presentation

Exercise format

There are many possible formats for developing oral communication skills, but this works well as an introductory exercise; alternatives are suggested below. If the students have been previously assigned to companies they can give their presentations individually within their company groups; the audience might consist of a second 'company' who would provide feedback for the speakers.

Introduction

The format of the introductory session depends on the students' previous experience of giving oral presentations. For relatively inexperienced students a discussion of what makes an effective presentation can be particularly sucessful. A simple method for such an introduction would be for the tutor to give a poor presentation lasting approximately five minutes. This might be badly structured, containing too much detail and using cluttered overheads that are changed too quickly, and it will therefore provide material for an open discussion. After this discussion the presentation could be repeated, using different overheads, in order to respond to the points brought up in the discussion.

After this introduction, the student handout should provide all of the information they need for the exercise.

Presentations

It is not practical to run more than about six presentations in one hour; if a large number of students are taking part in the exercise, several tutors are probably needed to cover all of the presentations. The tutor's role is to chair the delivery of the presentations and the feedback session, and provide written feedback for each student. Student assessors should make brief notes on all of the talks, but each one should have a specific talk on which they make more detailed notes and lead the feedback discussion. A handout is provided for this, which identifies criteria against which a talk might be assessed.

- Before starting, a timetable of speakers, titles and an order for the presentations should be prepared, and specific talks assigned to each of the student assessors.

- At the end of the session, the student assessors' feedback for their assigned presentation should be given, followed by a brief discussion (eg do the other students agree or disagree). It is essential to ensure that at least one good point emerges for all students. The tutor may want to summarise the feedback for each speaker, placing emphasis on the good points of their presentation and suggesting areas for improvement.

- If there is time after the presentations have been given, a general summary or discussion of some of the good and bad points that emerged overall could be included.

- Written feedback can be supplied using the tutor assessment form.

Features of the exercise

Most degree courses now require students to give oral presentations at some stage, but there are a number of reasons why this particular format works well:

a) Setting the talk in a specific context helps all of the students to take it seriously.

b) The commercial angle of the presentation helps students to see the (industrial) relevance of their chemistry.

c) Talking to an audience of about a dozen, with peer review, places them under a certain amount of pressure and most are pretty nervous, but even shy students seem to cope.

d) The peer review feedback has always been constructive and helpful to the speaker, and as assessors, students find out how to learn from what others do, both well and badly.

Adapting the exercise

Adapting the exercise is easy, through changes to the length, level, type of topic, scenario, or assessment, but points a-d above are worth bearing in mind.

Assessment

The tutor's assessment form is self-explanatory. Assessment forms can be filled in roughly during the talks and amended in the light of peer feedback. Individual forms can then be completed later, and an overall grade for the talk can be given, as indicated on the form.

Feedback from tutor

(Incorporating peer feedback)

Name of speaker:

_____

Title of presentation:

_____

Feedback	Excellent	Good	Average	Poor	Very poor
Did the presenter speak clearly?					
Was the talk well structured?					
Was your interest maintained?					
Was the science well explained?					
Were the visual aids good, and used to good effect?					

What were the strong points in the presentation?

_____

_____

_____

_____

_____

_____

Where might improvements be made?

_____

_____

_____

_____

_____

_____

Overall mark:
(A – excellent; B – good; C – average; D – poor; E – very poor)

9. Interviews and Interviewing

Summary

Outline of the exercise

In this exercise students are involved in both sides of a recruitment process. On the one hand, they construct the advertisement for a post within their company, interview a number of candidates and from these appoint someone for the job. And on the other, they must prepare a CV and covering letter for one of the advertised posts and take part in the selection procedure. The activity requires effective team work, and students gain insight into the problems faced by an interviewer. The experience gained in the exercise will undoubtedly be of use to students when they apply for jobs and are interviewed.

Key aims

■ to produce a good CV;

■ to develop good interview techniques; and

■ to develop team working skills.

Time requirements

■ 2.5 hours tutor contact time

■ Approximately 5–6 hours private study

■ Approximately 8 hours total student time

Timetable

The following timetable is suggested, and is most effective if the sessions are spread over one week.

1 hour	Introduction to CVs and interviews
6 hours	Student work:
	Construct adverts (0.5 hour)
	Prepare applications (4–5 hours)
	Plan the interviews (and review applications) (0.5 hour)
1 hour	Each interview session with 6 students requires a little over 1 hour
30 minutes	Debriefing

S9 Interviews and interviewing

Advertisement

You work for a company, which is a subsidiary of a multinational company called ACE. As part of the development of a new product, you have been given permission by ACE to employ an additional chemist. As the research committee it is up to you to decide what that post should be – a bench chemist ('loner' or team player), or a team leader (who might concentrate on developing ideas, checking out the literature for patents and synthetic methods, and guiding the group), a liaison chemist maintaining links between bench-chemists/plant-biologists/marketing, or another chemistry-related post you might identify.

You must construct the advertisement for this job. The salary will be £11–20,000, depending on the post and the experience of the appointee. The advertisement needs to

- give some background to the company (remember that advertisements are expensive, so it should be brief);

- give details of the post – for example, experience or qualifications needed;

- outline what the job will involve; and

- be eye-catching.

The advertisement should be no more than half-a-side of A4, and should be printed out as soon as possible, and displayed on a notice-board. Applications should require a CV, and a covering letter briefly explaining why the applicant thinks that they should get the job.

It may be useful to look in a magazine, for example *Chemistry in Britain*, to see how such advertisements are structured.

Interviewees

You are also looking for a job as you are worried by rumours that your company may be "restructuring". You must respond to an advertisement from a rival company (posted on the notice-board). It is essential that you 'sell yourself' in the best possible light, emphasising how your talents and experience make you the ideal candidate for the job. In your application (CV and covering letter), you should ensure that the specific skills you have developed (eg laboratory work, report writing, presentational skills, team work, academic ability and outside interests) are clearly identified and relate to the job advertised.

Your application must be submitted (clearly labelled with your name, and the company to whom you are applying) by the deadline given, so the interviewing panel has sufficient time to prepare for your interview.

Interviewers

Before interviewing any of the candidates for the post you have advertised, it is essential that, within the group, selection criteria for the post are agreed. Forms are provided for your comments and as a record of the decisions you have made. The appointment record form should be handed to the tutor at the end of the exercise.

The interviewing panel should prepare a timetable for interviews and for the subsequent debriefing/announcement and post it on the door of the interview room. The panel must decide how they want to run the interview (for example, they may or may not want to ask questions on technical chemistry problems or on leisure interests). Every member of the interview panel should be involved in the interview process, although two or three members might ask all of the questions at each interview.

Make sure the interviewee is relaxed – hard questions are fine, but do not reduce the candidate to tears! You must keep to time (so identify who will lead the interviewing); all of the candidates must be interviewed and a feedback session held in the time allocated. Decide how long each interview will be (remember that candidates might have questions) and allow a few minutes after each interview to jot down notes on the comments form provided, and prepare for the next applicant.

At the end of the interviews, you will need to decide whom to appoint. Assemble all of the candidates (at least 15 minutes before the end of the session if possible), and summarise the points that impressed you from the interviews – perhaps select the best CV, comment on the covering letters, and pick out strong points from the interviews themselves. Announce whom you are appointing at the end of the session.

Shortly after the interviews, it is a good idea to think about the way you conducted the interviews. In particular:

- Should you have spent more time planning the interviews?

- What sorts of questions were most informative?

- How might you conduct interviews better in future?

- Was there anything that really impressed you about the interviewees?

- Did the interviewees do anything particularly negative?

- What did you learn as an interviewer that should improve your chances of success when you are being interviewed?

There will be a final debriefing session with everyone present, to discuss the recruitment methods of the companies.

T9 Interviews and interviewing

One of the greatest benefits of this exercise is that, by carrying out interviews themselves, the students gain a better understanding of what an interviewer is hoping to learn.

Introduction

An introduction to writing effective CVs and good interview technique will be invaluable to students. If it is possible to involve experts from the University Careers Service, students will benefit.

The simplest format for the exercise is for all of the members of one company to apply for a single post. The student handout offers total flexibility in the job description, so some discussion will be needed for groups to decide on the position that they will advertise and the selection criteria for candidates. If time is short, it may be necessary to outline the requirements for the job in advance. Three scenarios are suggested:

i) Website editor

In order to raise their profile the company has decided to launch a website and have already employed an IT expert to set up the page. They now want to recruit an editor for the website. No knowledge of the technical side of the Internet is required as this is dealt with by the IT expert. The candidates must have a solid background in chemistry, and a wider interest in communicating chemistry, often to a non-technical audience.

ii) Marketing

This role is for someone with a chemical background who will liaise with customers and chemists or other scientists working on site. The candidate must therefore be able to understand the technical nature of the company's research and development, and must be able to make this relevant to customers. They must also have the ability to translate a customer's needs into projects that the R&D department can develop. This job requires someone who is technically very able, with commercial awareness and an ability to communicate scientific principles clearly.

iii) Practical Chemist

The students must prepare/receive a technical interview; the panel might also look for good practical skills and a professional approach to laboratory work in the candidates. The job description for this role will need to be fairly specific to allow for some revision by the candidates.

The interviews

Practicalities depend on the number of students participating, and the number of interviews that can be run in parallel. Interviews should be between eight and twelve minutes long, with two to three minutes between each one. Organise the companies so that they only meet each other once – *ie* avoid them acting as both interviewers and applicants to the same company.

A tutor should be present at each set of interviews in order to make sure that the students are properly organised before they start; for example, to ensure that there is an interview timetable on the door, a chair outside for waiting interviewees, that the room is laid out for the interviews, and roles have been allocated (*eg* who will lead the questions). The tutor should also ensure that the interviews keep to time. Although the tutor is not essential, his/her (silent) presence helps students to take the exercise more seriously.

A form is provided which is similar to one usually filled in by interviewers and can be used to help panels make their final decision, especially if two or three candidates are very close.

Debriefing
A final feedback session is useful, involving all of the students, someone from the university Careers Service and all of the tutors who have observed interviews. In open discussion, the students will probably identify several important points, and the observers/careers advisors will almost certainly have valuable feedback.

Features of this exercise
In this exercise students sit on both sides of the table and might learn more from conducting an interview than from having one. They understand why their colleagues do well or badly, realise why certain questions are effective, and develop the types of answer that might impress interviewers. Input from the University Careers Service during this exercise is particularly effective.

Adapting/extending the exercise
It is easy to modify the basic exercise, but changes may lead to more time being required, or special facilities. Suggestions include:

■ Twenty minute interviews, requiring at least five minutes of technical questions

■ Video the candidates, so they can see their own performance and discuss it afterwards

Assessment

It is simple and effective to provide informal feedback from the panel to the candidates using a comment form, and panels should be asked to identify at least one good area and one area for improvement for each candidate. A mark could be generated by assessing the following (possible maximum marks are given):

CV	20 marks
Letter of application	5 marks
Performance at interview	20 marks
Contribution to the interviewing process	10 marks
The panel's appointment record form	5 marks
Total	60 marks

Appointment Record Form

Job Title:

_____

Number of Applicants:

_____

Criteria for Shortlist:

_____

_____

Final appointment based on:

CV	Letter of application	Presentation	Informal Interview	Formal Interview	Second Interview	Other*

*If 'other', explain:

Date(s) of interview(s):

_____

Insert around six main criteria for the job (e.g. technical ability, initiative, or reliability) in the table below. At the end of each interview, insert your assessment of the candidate against the criteria (A: outstanding; B: excellent; C: very good; D: quite good; E: poor; X: unsatisfactory), and assign an overall assessment (the criteria need not carry equal weight).

Candidate's names	Criteria						Overall assessment
	1	2	3	4	5	6	
1							
2							
3							
4							
5							
6							

Decision:
1st choice

_____

2nd choice

_____

Reasons for final choice:

_____

_____

Names of interview panel:

1)   (Chairperson)   2)   3)

4)   5)   6)

Signed: _____   Date: _____

Submit this form with a copy of the advertisement attached.

10. Poster Presentation

Summary

Outline of the exercise

Students work in groups to prepare a poster presentation. In the scenario outlined in this book, the students represent key members of a research committee, who are hoping to receive funding for their latest ideas. Posters must therefore include information on the chemistry behind the idea, its commercial applicability and the cost of the project. Students are encouraged to search recent literature for suitable topics for their posters. The posters are presented to a panel of judges, some of whom might be external to the department, and "funding is awarded" to the group with the best poster.

Key aims

- to describe a piece of chemistry using a high quality poster;

- to develop skills in information retrieval, written delivery, visual delivery, and team work.

Time requirements

- 2 hours tutor contact time

- 10 hours private study

- 12 hours total student time

Timetable

The following timetable is suggested and is most effective if spread over a one-week period.

20 mins	Introduction
10 hours	Students prepare their posters
1.5 hours	Judging of the poster display (1 hour, with one or two students from each company present)
	Final debriefing and prize giving (all students present, 20 minutes)

S10 Poster presentation

ACE, the parent company you work for, would like to demonstrate their confidence in you by asking you to outline a new area of research of your choice, for which they might fund an initial R&D programme of £200,000 over 3 years. Several other subsidiaries of ACE have also been asked to put forward a proposal, but only one will be funded.

ACE is keen to diversify and is therefore willing to consider any chemistry research proposal (eg new materials, pharmaceuticals or agrochemicals). If specialist work is required (for example, testing a new drug), external companies can be contracted. However, the main thrust of the chemistry must be conducted in-house. The cost of an employee to a company is about double their salary (approximately £120,000 to fund a reasonably qualified chemist for three years).

Sources like *Chemistry in Britain*, the science pages of national newspapers, or journals such as *Chemical Communications* should help you to identify a topic that really interests you, has commercial potential, and for which you could envisage a worthwhile research programme.

ACE has asked all applicants to prepare a poster that will be assessed by a high powered team from ACE including the Chairman (who does not have a chemistry background) and some of the top research chemists. They will be strongly influenced by a clear presentation of why the topic is important, what new development has taken place, and how it might be further developed in the future. Your planned research programme will probably be quite brief at this stage – an outline of the work you would carry out, the resources needed, and the aims of the project. You need to set up meetings of your company, at which you:

■ decide how to choose your topic

■ choose your topic;

■ plan the poster;

■ agree who will do what (*eg* writing, preparing graphics, finding materials, following up references); and

■ prepare the poster.

The best posters are clear and easy to follow, have visual impact, and do not contain too much information. In this case an overview for the non-specialist and some further details for specialist chemists (including several references to the chemical literature) are needed. The poster must have well defined sections; for example, background, the major recent development, an outline of what you plan to do and achieve (maybe including rough cost), and a summary. It must also contain a title, the names of the R&D team and your company's name. The whole presentation must fit within 1 m x 1 m, but should at least fill a 90 x 60 cm area, and should be readable at a distance of about 1 m. Coloured paper, pictures, graphics and any other materials which would enhance the quality of your poster can be used.

T10 Poster presentation

The tutor input for this exercise is simply:

■ distributing the information sheet;

■ arranging for suitable resources to be available for the students to prepare posters;

■ providing an area for the posters to be displayed; and

■ inviting judges to assess the posters.

To produce a reasonable poster, the team members need to contribute approximately 10 hours work each. Three to four days is an adequate timescale and provides enough time whilst instilling a sense of urgency, although it is also possible to run the exercise as a one-day event. If the timescale is extended beyond a few days, it can take over from other course commitments. The size of the poster ($1m^2$) is based on a typical conference poster.

The success of the event is driven by having the posters on display for the whole department, and using external judges for the assessment. This is a good opportunity to invite colleagues from industry to see the undergraduates' work. Awarding prizes for the best poster(s) at the end of the event will ensure that it finishes on a high note; if this is the last exercise of a module on communicating chemistry, it might also be appropriate to reward good performances from exercises throughout the course.

Adapting the exercise

A (fictitious) reason for requesting a poster can help the exercise. The format described here builds on previous exercises in the module, but it would also run well as a stand-alone exercise, or with a modified scenario.

Assessment

An assessment form is provided. Four criteria are given for the assessment of the posters, although a wider range could be used. One or two students from each company can be asked to attend their poster when the judges are assessing it.

Criteria for marking:

- Choice of topic   [10]
- Visual Impact – use of layout, graphics, colour   [20]
- Clarity – was the poster easy to follow?   [20]
- Scientific accuracy   [20]
- Commercial relevance – was the case well made?   [10]
- References   [10]
- Discussion with company   [10]

Each category is marked out of 10.

	Company Name			
	1	**2**	**3**	**4**
Choice of subject				
Visual impact				
Clarity				
Scientific accuracy				
Commercial relevance				
References				
Discussion with company				
Overall				

References

Chapter 1 – The Fluorofen Problem
P. D. Bailey, Coaxing Chemists to Communicate, *U. Chem. Ed.*, 1997, **1**, 31–36.

Chapter 2 – Scientific Paper
Patrick Bailey, reported by Billy Kerr, The Appliance of Science, *Proceedings of Variety in Chemistry Teaching 1993*, pp25–31, (Mary Aitken, ed.), The Royal Society of Chemistry, 1993.

C. J. Garratt and T. L. Overton, Scientific Papers as a Teaching Aid, *Educ. Chem.*, 1996, **33**, 137–138.

Chaper 4 – World Wide Web Treasure Hunt
Pat Bailey and Sara Shinton, with report by Simon Higgins, 'Teaching Communication Skills in Chemistry', *Proceedings of Variety in Chemistry Teaching 1997*, pp36–7, (John Garratt and Tina Overton, eds.), The Royal Society of Chemistry, 1997.

Chaper 7 – Hwuche-Hwuche Bark
Pat Bailey, with report by Jane Tomlinson, 'Hwuche-Hwuche Bark; Applying Chemical Knowledge', *Proceedings of Variety in Chemistry Teaching 1996*, p31, (John Garratt and Tina Overton, eds.), The Royal Society of Chemistry, 1996.

Appendix A

Reasonable answers to handouts from Exercise 2

Section 1 Introduction

This section is primarily used to familiarise the students with the structures and reactions in the paper. There are quite a few terms in the introduction to the paper which they might not be familiar with. It is perhaps surprising that there are very few terms for which, even if the students are unfamiliar with them, the meaning of the paper is lost.

1. **What are alkaloids?**
   Answer: Natural products that contain a basic nitrogen.

2. **Why should anyone want to synthesise alkaloids?**
   Answer: Alkaloids are usually toxic. If used in controlled amounts, or related analogues are prepared, they can be used as medicines.

3. **What features of a synthesis would make it a 'good' synthesis?**
   Answer: Short; efficient (high yield); cheap (reagents and conditions); environmentally friendly.

4. **Why might an understanding of the mechanism of the Pictet-Spengler reaction help to achieve a 'good' synthesis of an indolic alkaloid?**
   Answer: Based on the mechanism, it might be possible to modify the conditions to improve the yield and/or the stereo-control.

Section 2 Methods for following reaction pathways

This is one of the most important and interactive sessions – a chance for the students to realise how many techniques they know about. Most physical methods could provide mechanistic feedback, and the tutor needs to summarise clearly the general and specific techniques for tackling the problem in the paper.

1. **Does the author give you the impression that route a is more likely than route b, or vice versa, or that either route is equally likely, or that both routes occur simultaneously? From the way the author has presented the case, suggest likely odds from a bookmaker for the Pictet-Spengler reaction proceeding via route a or route b.**

   Answer: The author produces a fairly balanced case, which is why the experimental research was conducted. In other words, both pathways are quite likely, and they might be occurring simultaneously (although it's unlikely). Then state that, almost certainly, one of the pathways is followed, and take a vote on it – typically a:b is *ca* 2:1.

2. **Many different methods have been used to study reaction pathways; list as many as you can (general methods and specific techniques). Decide whether these techniques could be used to investigate which the two suggested routes for the Pictet-Spengler reaction is correct.**

   Answer: Here is a selection of suggestions, divided into methods (*ie* the general tactic used to try to discover the mechanism) and associated techniques (practical ways of tackling the method):

List of Methods	List of Techniques
Isolate an intermediate	NMR (^{1}H, ^{13}C, others)[†]
Identify by-products*	IR spectroscopy
Detect a (transient) intermediate	Ultraviolet spectroscopy
Try reactions with related	Mass spectroscopy
compounds, especially if they	ESR
are designed to test a proposed	Low Temperature/matrix isolation
mechanism*	Radioactivity detect: ^{3}H,
Monitor the fate of labelled	^{14}C
atoms	Use spectroscopic techniques or
Study the kinetics/	chromatography (eg GC, HPLC)
thermodynamics	to monitor progress of reaction

Students should suggest plenty of techniques and methods. If time permits, more extensive discussion could be included (eg discussing how NMR/IR/UV provide successively less detail, but can detect shorter lived intermediates; or the types of analogues that one might make to test proposed mechanisms).

Virtually all of the methods and techniques could be used for studying the Pictet-Spengler reaction, but if pathways a and b need differentiating, the methods marked * and the techniques marked † are the most likely to provide answers.

Section 3 Examining the author's strategy

This is another particularly important section – the students must grasp the answer to question 3, or the whole point of the subsequent experiments is lost.

1. **What general technique did the author actually use?**
    Preparation of analogues (and isotopic labelling)

2. **Draw structures of the intermediates you would expect to obtain in the formation of 2,3-dimethyl-1,2,3,4-tetrahydro-3-aza-β-carboline (7) if the reaction proceeds via:**

    (i)   **route a (*ie* the 3-aza-analogue of 3);**

Position of label

(ii) route b (*ie* the 3-aza-analogue of 4)

Position of label

There is something special about the intermediate from (ii), which might allow the pathways to be distinguished – what is it? (NB You might find this much easier to see if you make a simple model of the intermediate).

Answer: Ignoring any label, the positions marked ▲ and * are identical.

3. **Hence, explain why the author expected to be able to clarify the mechanism by carrying out the reaction using labelled methanal, and then studying the distribution of label in the final product.**

   Answer: The key feature of structure (ii) from pathway b is that carbons ▲ and * occupy equivalent positions and are equally likely to migrate in the next step; this would leave the isotope equally distributed between two positions (1 and 4) in the final product (**7**). If pathway a were followed, intermediate (i) would lead to product in which the label is found only on position 1 in the product (**7**).

4. **Formaldehyde could be labelled with ^{2}H, ^{3}H, ^{13}C, ^{14}C, ^{17}O, or ^{18}O. No labelled formaldehyde was commercially available at the time this experiment was done, so the author had to make some. Consider each of the isotopes and decide how you would determine their positions in the final product. Which do you think would be the best to use?**

   Answer: *Positions of labels*
   ^{13}C from intense peak(s) in the ^{13}C NMR
   ^{2}H by disappearance of signal(s) in ^{1}H NMR
   ^{3}H, ^{14}C, ^{18}O by identification of radioactive products after chemical degradation

   Answer: *Easiest method*
   ^{13}C should give the most unambiguous results, but ^{2}H would be fine too. Radioactive isotopes are trickier to handle, and provide answers less easily.

Section 4 The synthesis of products

Sections 4–6 can be covered quite quickly, despite the fact that there are quite a few questions to be answered. If the tutor addresses the whole class every two to three minutes, getting feedback to a few of the questions at a time, rather than having longer plenary sessions, the pace is kept up and groups do not get behind. If possible, finish on a high note – for example, emphasising that the students themselves have come up with viable solutions to a difficult research problem, and that the mechanistic results in the paper have actually led to a better synthetic route to some important natural products.

1. **Which isotope did the author opt for? Why do you think this isotope was chosen?**
   Which isotope? ^{2}H
   Why?   Easy to handle, should give clear results directly from NMR; cheaper than ^{13}C.

2. **Why do you think the author initially carried out the synthesis in the absence of isotope?**

   Answer: Isotope labelling is a costly procedure. The author would have conducted earlier non-labelled experiments to develop high yielding conditions; moreover, the non-labelled product was needed for assignment of all of the NMR signals, for comparison with the labelled product.

3. **What is the origin of compound 8 which was found in 50% yield? (In other words, what are the sources of the tricycle X, the CH₂ Y and the MeO Z in structure 8?)**

   This is formed by the further reaction of the desired product with methanol and formaldehyde. **X** is from compound **7**, **Y** is from CH₂O, and **Z** is from MeOH.

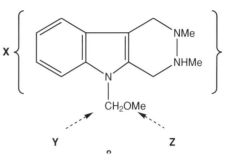

4. **The author is apparently satisfied with a yield of the desired product (7) of only 26%. Why?**

   As the by-product is formed from the desired product (7), it must have been formed via the same mechanism. As it is the mechanism that the author is studying, he has achieved a 76% yield of products following this pathway and, provided there is enough product to analyse, the yield is acceptable.

5. **Decide which technique(s) you would use to investigate the isotopic composition of compound 7 and give reasons for your choice.**

   The obvious choices are NMR, comparing labelled and unlabelled products, and mass spectrometry where the molecular ion would be at M+2 (for CD₂O) or M+1 (for ¹³C). Mass spectrometry would ensure that the label had been incorporated into the molecule successfully, but would probably not indicate where it was located.

Section 5 Analysis of results

1. **Explain how you might expect to be able to use ¹H NMR to tell whether the product had deuterium atoms at positions C(1) and/or C(4)? (In other words, what specific change in the NMR of the labelled product versus the unlabelled product should allow the pathway to be elucidated?)**

   From answers to earlier questions, it should be clear that the presence of deuterium will depress both peaks to half intensity if route b is followed (D₂ at positions 1 or 4), and eliminate one peak entirely if route a dominates (D₂ at position 1 only).

2. **The C(1) and C(4) protons are broad and poorly resolved in the ¹H NMR spectrum of the undeuterated adduct. Why can these signals not be used to determine the location of deuterium in the deuterated adduct?**

   The two possible locations cannot be differentiated in this way as the peaks cannot be resolved.

3. **The N-methyl hydrogens resonate in the region around δ2.5. Explain why the two methyl groups give signals at slightly different chemical shifts, and why both signals were present as sharp singlets in the undeuterated adduct.**

   The methyl groups have different chemical environments, leading to signals at different chemical shifts. The singlet peaks indicate there are no adjacent proton(s) to which the methyl protons are coupled.

4. **Do you agree that there are four signals in the ¹H NMR for each of the N-methyl groups?**

5. **The deuterated compound gave rise to a spectrum containing 'four singlets of roughly equal intensity in the region of δ2.5'. Look at the structure below representing the deuterated product, and decide what arrangement of hydrogen and deuterium in the positions X and Y could give rise to this observation. (Note that the ratio of CH₂/CD₂ remains intact; isotopes cause slight changes in chemical shifts).**

   As the formaldehyde is fully deuterated, and the starting material contains a CH₂ group, positions 1 and 4 in the product (7) must have CH₂ or CD₂ groups. The four compounds therefore must be:

   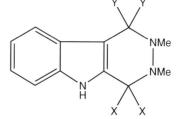

   a) X = D, Y = H   c) X = H, Y = H
   b) X = H, Y = D   d) X = D, Y = D

6. **Suggest the identities of the molecules which give rise to the parent ions which appear at m/z 201, 203 and 205 in the spectrum.**

   m/z 201   Product 7
   m/z 203   Product 7 containing 2 deuterium atoms
   m/z 205   Product 7 containing 4 deuterium atoms

7. **Does your interpretation of the mass spectrum support your ideas about the identity of the deuterated product that you formulated from the NMR spectrum?**

   Yes: m/z 201; compound c) from Q5
              m/z 203; compounds a) and b) from Q5
              m/z 205; compound d) from Q5

Section 6 The author's conclusions

1. **Did you reach the same conclusion as the author? (If not, who's right?)**

2. **Are these results consistent with pathway a, or pathway b, or neither?**
   Neither!

3. **The most unexpected observation was the formation of a product containing four deuterium atoms – convince yourself that Scheme 2 would allow this to occur, and hence that all of the products 9–12 could have been formed if the mechanism in Scheme 2 were operating.**

Assessment

■ Students could be asked to summarise, in their own words (maximum of fifty), what the author claims to have discovered about the pathway taken by the Pictet-Spengler reaction.

Appendix B – Exercise 6

Topic 1
Microwave Synthesis

Title/keyword	microwave	microwave	microwave
Journal title	*Nature*	*Chemical Society*	*Chemical Communications*
From/to	1988/1988	1988/1988	1996/1996
Document type	All	All	All
Number of hits	6 hits	27	6 hits
Comments	Only one with DRB	One is clearly the same group as reference 1	1 hit with authors W and B.

Topic 2
$Rh_4(CO)_{12}$

References 1 and 2 found through *Chemical Abstracts*. Author names given as clues.

Title/keyword	rhodium + cataly* + selective synthesis
Address	Japan
From/to	1991/1991
Document type	All
Number of hits	2 hits

Topic 3
Anatoxin

Reference 1 through *Chemical Abstracts*. Author name given as clue.

Title/keyword	Anatoxin	Anatoxin
Author		Skrinjar
Journal title	*Pharmacology*, vol. 135	
From/to	1987/1987	1992/1992
Document type	All	All
Number of hits	1 hit	1 hit

Topic 4
Biological Applications of Poly(diacetylenes)

Title/keyword	Influenza		
Author		Charych_*	Charych_*
Journal title	american+chemical	american+chemical	medic*
From/to	1993/1993	1995/1995	1996/1997
Number of hits	4 hits	3 hits	1 hit
Comments	1 hit with author DHC.	1 hit with JACS.	

Topic 5
Main-chain Thermotropic Liquid Crystalline Polymers

References 1 and 2 found through *Chemical Abstracts*. Author names (1) Rovielli and Sirigu, (2) Jackson and Kuhfuss.

Title/keyword	liquid + crystal* + polymer* + thermotropic
Journal title	polymer + engineering
From/to	1987/1987
Number of hits	4 hits
Comments	1 hit with a single author.

Topic 6
Anisotropic Gels

Title/keyword	anisotropic+gel*	anisotropic	
Author		Kim	Hikmet_RAM
Journal title	physic*	journal of applied physics	journal of applied physics
From/to	1990/1990	1992/1992	1990/1993
Number of hits	1 hit	2 hits	2 hits
Comments		Reference should be easily identified.	1 hit with 2 authors.

Topic 7
Photodissociation of methane

Reference 1 found through *Chemical Abstracts*. Author name is Mahon.

Title/Keyword	CH_4 (or photodissociat*)	CH_4
Author	Black	Morley
Journal title		journal of chemical physics
From/to	1982/1982	1990/1997
Number of hits	1 hit	1 hit

Topic 8
Cyclobutadiene

All references found through *Chemical Abstracts*. Author names (1) Longuett-Higgins, (2) Emerson, (3) Pebit.

Topic 9
Vinyl Benzoate Polymerisation

All references found through *Chemical Abstracts*. Author names (1) Morrison, Gleason and Stannett, (2) Adelman, (3) Vrancken and Smets.

Topic 10
Suaveoline

Reference 1 found through *Chemical Abstracts*. Author name is Majumdar.

Title/Keyword		suaveoline
Author	Endress	
Journal title	phytochemistry	
From/to	1990/1997	1993/1997
Number of hits	3 hits	8 hits
Comments	2 hits for Endress.	1 hit authors B and M.

Topic 11
Odour-Structure Relationships

References 1 and 2 found through *Chemical Abstracts*. Author names (1) Amoore, (2) Russell and Hills.

Title/Keyword	odor
Journal title	pharmaceut*
From/to	1992/1992
Number of hits	1 hit

Topic 12
FR-900848

Title/Keyword	antifungal, fungal, FR900848	antifungal, fungal, FR900848	antifungal, fungal, FR900848
Author	Yoshida_M*	Barrett	Barrett
From/to	1990/1990	1995/1995	1996/1996
Number of hits	2 hits	6 hits	7 hits
Comments	Reference should be easily identified.	Only one synthetic with AJPW as a co-author	One obvious

Topic 13
Cis-Platin

Reference 1 is found through *Chemical Abstracts*. First author name is Rosenburg.

Title/Keyword	cis-platin, anticancer, cancer	
Author	Lippard	Lippard
Journal title		inorganic chemistry
From/to	1981/1989	1990/1992
Document type		review
Number of hits	2 hits	1 hit
Comments	1 hit with 3 other authors.	

Topic 14
Polymer Dispersed Liquid Crystals

Title/Keyword			polymer dispersed liquid crystal*
Author	doane	amundson	
Journal title	physic*	chemi*	polymer
From/to	1986/1986	1994/1994	1992/1992
Number of hits	2 hits	1 hit	2 hits
Comments	1 hit with 1 other author.		1 hit with ZZZ as author.

Topic 15
Danishefsky's Diene

Reference 1 is found through *Chemical Abstracts*. Author name is in clue.

Title/keyword		diene*
Author	Danishefsky	Yamamoto
Journal title	american chemical society	american chemical society
From/to	1982/1982	1990/1997
Number of hits	4 hits	3 hits
Comments	1 hits with JFK and SK.	1 hit mentions imines.

Topic 16
Reversible Oxygen Carriers

Reference 1 is found through *Chemical Abstracts*. First author name is Collman.

Title/keyword		oxygen, dioxygen
Author	busch	
Journal title	american	macro*, polymer*
From/to	1983/1983	1988/1988
Number of hits	13 hits	26 hits
Comments	One hit with 4 authors should be obvious.	1 hit with MO and HN as authors.

Topic 17
Liquid Crystalline Metal Complexes

Title/keyword		Vanadium + Nickel	metallomeso*
Author	Nolte_RJM	Hoshino	
From/to	1988/1988	1991/1991	1993/1993
Number of hits	11 hits	2 hits	14 hits
Comments	3 hits in a Dutch journal – 1 obvious.		2 hits in a German journal, author PE – 1 obvious.

Topic 18
Electrospray Ionisation

Title/keyword	Electrospray	Electrospray	
Author	Yamashita	Emmett	Sweetman
Journal title			clinical
From / to	1984/1984	1994/1994	1996/1996
Number of hits	2 hits	2 hits	1 hit
Comments	Earlier paper introduces subject.	1 hit with second author RMC.	

Topic 19
Comets and the origins of life

Title/Keyword	chem* + atmosphere	comets + life	org*
Author			McKay
Journal title	Nature	space	Science
From/to	1983/1987	1993/1997	1997/
Number of hits	4 hits	7 hits	5 hits
Comments	1 with correct initials	1 with correct initials	1 with correct initials

Topic 20
Ionic liquids

Title/Keyword	ionic liquids	ionic liquids	ionic liquids
Author			
Journal title	inorganic	tech*	communication*
From/to	1981/1984	1990s	1997/1999
Number of hits	6 hits	3 hits	8 hits
Comments	1 with correct initials	1 with correct initials	1 with 5 authors

Topic 21
Transition Metal σ-complexes

Title/keyword		transition metal*	transition metal* + CH$_4$, transition metal* + methane
Author	Swansen		
Journal title	american	journal of the american chemical society	journal of the american chemical society
From/to	1984/1984	1984/1984	1995/1995
Number of hits	8 hits	33 hits	6 hits
Comments	2 hits in JACS, 1 should be obvious.	1 hit with authors JYS and RH.	1 hit with author Burns.

Topic 22
The Isomerisation of Carboranes

Reference 1 from *Chemical Abstracts*. Author names in clue.

Title/keyword	Carboranes	carboranes + mechani*
Author	Wales_D	
From/to	1993/1993	1993/1993
Number of hits	1 hit	10 hits
Comments		1 hit with author SD.

Topic 23
Nitric Oxide

Title/keyword	nitric oxide	nitric oxide	nitric oxide
Author	Moncada		Rossaint
Journal title		chem*	
From/to	1987/1987	1993/1993	1996
Document type		review	
Number of hits	7 hits	3 hits	3 hits
Comments	1 hit with AGF as co-author.	Only 1 hit with authors ARB and DLHW	1 hit in an enzyme related journal.

Topic 24
Taxol

Reference 1 found through *Chemical Abstracts*. Author name in clue.

Title/keyword	taxol	taxol
Author	Nicolaou	Holton
From/to	1993/1997	1994/1994
Document type	review	
Number of hits	3 hits	4 hits
Comments	One of these is obvious.	2 papers describe the first total synthesis.

Topic 25
Organometallic Complexes of C_{60}

BIDS will not allow searching by chemical formula, so students should use *Chemical Abstracts* or determine key words from the clues given.

Title/keyword	organometallic + C_{60}	Iridium + C_{60}	C_{60} + arene
From/to	1981/1997	1981/1997	1990/1997
Number of hits	21 hits	7 hits	8 hits
Comments	5 hits when specifying Platinum, of which the *Science* article would provide the best introduction.	2 hits in widely available journals. Reading the abstracts of these will identify the relevant paper.	Of those in widely available journals, the abstracts will identify the suggested reference.

Topic 26
Symmetric Chlorine Dioxide

Reference 1 found through *Chemical Abstracts*. Authors names are Derby and Hutchinson.

Title/keyword	chlorine dioxide	chlorine dioxide, OClO, ClO_2
Journal title	chem* + phys*	
From/to	1994/1994	1987/1987
Number of hits	8 hits	31 hits
Comments	1 hit with SH and JBN.	1 hit with authors SS, GHM and RWS.

Topic 27
Dendrimers

Title/keyword	Dendrimer*		Arborols + cascade molecule*
Author		Hawker_CJ	
From/to	1990/1990	1990/1990	
Number of hits	6 hits	2 hits	8 hits
Comments	3 hits with DAT. 1 major paper obvious.	1 hit describes a new convergent approach.	1 hit with HBM as author.

Topic 28
Ascaridole

All references should be found using *Chemical Abstracts*. First author names (1) given in clue, (2) Schenck, (3) Srininvasan, with co-authors Brown and White.

Topic 29
Artemisinin

Reference 2 should be found using *Chemical Abstracts*. Author name is Qinghausu Research Group.

Title/keyword	Artemisinin	Artemisinin + synthesis
Author	Butler	
Journal title		american + chemical
From/to	1990/1997	1990/1997
Number of hits	1 hit	6 hits
Comments		2 hits with stereoselective syntheses, one authored by MAA + 2 others.

Topic 30
The Wacker Process

References 1 and 2 should be found using *Chemical Abstracts*. Author names (1) given in clue, (2) Smidt

Title/keyword	Palladium
Address	Japan
From/to	1983/1987
Document type	review
Number of hits	6 hits
Comments	Only 1 hit refers to olefins.

The references

Topic 1
Microwave synthesis

D. R. Baghurst *et al.*, *Nature*, 1988, **332**, 311.

D. R. Baghurst *et al.*, *J. Chem. Soc., Chem. Commun.*, 1988, 829.

C. G. Wu and T. Bein, *J. Chem. Soc., Chem. Commun.*, 1996, 925.

Topic 2
$Rh_4(CO)_{12}$

B. L. Booth *et al.*, *J. Organomet. Chem.*, 1968, **14**, 417.

C. H. Wei, *Inorg. Chem.*, 1969, **8**, 2384.

T. Joh et al., *Organometallics*, 1991, **10**, 2493

Topic 3
Anatoxin

J. P. Devlin *et al.*, *Can. J. Chem.*, 1977, **55**, 1367.

X. Zhang *et al.*, *Eur. J. Pharmacol.*, 1987, **135**, 457.

T. Joh *et al.*, *Tetrahedron Asymmetry*, 1992, **3**, 1263.

Topic 4
Biological Applications of Poly(diacetylenes)

D. H. Charych *et al.*, *J. Am. Chem. Soc.*, 1993, **115**, 1146.

D. H. Charych *et al.*, *J. Am. Chem. Soc.*, 1995, **117**, 829.

D. H. Charych *et al.*, *J. Med. Chem.*, 1996, **39**, 1018.

Topic 5
Main-chain Thermotropic Liquid Crystalline Polymers

A. Roviello and A. Sirigu, *J. Poly. Sci.: Poly. Lett. Ed.*, 1975, **13**, 455.

W. J. Jackson and H. F. Kuhfuss, *J. Poly. Sci.: Poly. Chem. Ed.*, 1976, **14**, 2043.

G. Kiss, *Polym. Eng. Sci.*, 1987, **27**, 410.

Topic 6
Anisotropic Gels

R. A. M. Hikmet, *J. Appl. Phys.*, 1990, **68**, 4406.

A Jakli *et al.*, *J. Appl. Phys.*, 1992, **72**, 3161.

R. A. M. Hikmet and C. de Witz, *J. Appl. Phys.*, 1991, **70**, 1265.

Topic 7
Photodissociation of methane

B. H. Mahon *et al.*, *J. Chem. Phys.*, 1962, **37**, 207.

T. G. Slanger *et al.*, *J. Chem. Phys.*, 1982, **77**, 2432.

D. H. Mourdout *et al.*, *J. Chem. Phys.*, 1993, **98**, 2054.

Topic 8
Cyclobutadiene

H. C. Longuet-Higgins *et al.*, *J. Chem. Soc.*, 1956, 1969.

G. F. Emerson et al., *J. Am. Chem. Soc.*, 1965, **87**, 131.

R. Pebit , *J. Organomet. Chem.*, 1975, **100**, 205.

Topic 9
Vinyl Benzoate Polymerisation

E.D. Morrison *et al*, *J. Poly. Sci.*, 1959, **36**, 267.

R.L. Adelman, *J. Org. Chem.*, 1949, **14**, 1057.

A. Vrancken and G. Smets, *Makromol. Chem.*, 1959, **30**, 197.

Topic 10
Suaveoline

S. P. Majumber *et al.*, *Tetrahedron Lett.*, 1972, 1563.

S. Enders *et al.*, *Phytochemistry*, 1993, **32**, 725. (referred to as ENDRESS on BIDS)

P. D. Bailey *et al.*, *J. Chem. Soc., Chem. Commun.*, 1996, 1479.

Topic 11
Odour-Structure Relationships

J. E. Amoore, *Nature*, 1967, **214**, 1095.

G. F. Russell *et al.*, *Science*, 1971, **172**, 1043.

A. Tsantili-Kakoulidou *et al.*, *Pharmaceutical Research*, 1992, **9**, 1321.

Topic 12
FR-900848

M. Yoshida *et al*, *J. Antibiot.*, 1990, **43**, 748.

A. G. M. Barrett *et al.*, *J. Chem. Soc., Chem. Commun.*, 1995, 649.

A. G. M. Barrett *et al.*, *J. Am. Chem. Soc.*, 1996, **118**, 11030.

Topic 13
Cis-Platin

B. Rosenberg *et al.*, *Nature*, 1969, **222**, 385.

S. E. Sherman *et al.*, *Science*, 1985, **230**, 412.

S. L. Bruhn *et al.*, *Prog. Inorg. Chem.*, 1990, **38**, 477.

Topic 14
Polymer Dispersed Liquid Crystals.

J. W. Doane *et al.*, *Appl. Phys. Lett.*, 1986, **48**, 269.

A. J. Lovinger *et al.*, *Chem. Mater.*, 1994, **6**, 1726.

Z. Z. Zhong *et al.*, *J. Poly. Sci., Part B: Poly. Phys.*, 1992, **30**, 1443.

Topic 15
Danishefsky's Diene

S. Danishefsky *et al.*, *J. Am. Chem. Soc.*, 1974, **96**, 7807.

S. Danishefsky *et al.*, *J. Am. Chem. Soc.*, 1982, **104**, 358.

K. Ishihara *et al.*, *J. Am. Chem. Soc.*, 1994, **116**, 10520.

Topic 16
Reversible Oxygen Carriers

J. P. Collman *et al.*, *J. Am. Chem. Soc.*, 1973, **95**, 7868.

D. H. Busch *et al.*, *J. Am. Chem. Soc.*, 1983, **105**, 298.

E. Tsuchida *et al.*, *Macromolecules*, 1988, **21**, 1590.

Topic 17
Liquid Crystalline Metal Complexes

R. J. M. Nolte *et al.*, *Recl. Trav. Chim. Pays-Bas*, 1988, **107**, 615.

N. Hoshino *et al.*, *Inorg. Chem.*, 1991, **30**, 3091.

P. Espinet *et al.*, *Angew. Chem., Intl. Ed. Engl.*, 1993, **32**, 1201.

Topic 18
Electrospray Ionisation

M. Yamashita and J. B. Fenn, *J. Phys. Chem.*, 1984, **88**, 4451.

M. R. Emmett *et al.*, *J. Am. Soc. Mass Spectrom.*, 1994, **5**, 605.

L. Sweetman *et al.*, *Clinical Chem.*, 1996, **42**, 345.

Topic 19
Comets and the Origins of Life

B. Fegley, R. G. Prinn, H. Hartman and G. H. Watkins, *Nature*, 1986, **319**, 305;

J. Oro, T. Mills and A. Lazcano, *Advances in Space Research*, 1994, **15**, 81;

C. P. McKay and W. J. Borucki, *Science*, 1997, **276**, 390.

Topic 20
Ionic Liquids

J. S. Wilkes, J. A. Levisky, R. A. Wilson and C. L. Hussey, *Inorg. Chem.*, 1982, **21**, 1263;

K. R. Seddon, *J. Chem. Tech. and Biotech.*, 1997, **68**, 351;

J. G. Huddleston, H. D. Willauer, R. P. Swatloski, A. E. Visser and R. D.Rogers, *Chem. Commun.*, 1998, 1765.

Topic 21
Transition Metal σ-complexes

G. J. Kubas *et al.*, *J. Am. Chem. Soc.*, 1984, **106**, 451.

J.-Y. Saillard *et al.*, *J. Am. Chem. Soc.*, 1984, **106**, 2006.

X.-L. Luo *et al.*, *J. Am. Chem. Soc.*, 1995, **117**, 1159.

Topic 22
The Isomerisation of Carboranes

D. Grafstein *et al.*, *Inorg. Chem.*, 1963, **2**, 1128.

D. J. Wales *et al.*, *J. Am. Chem. Soc.*, 1993, **115**, 1557.

S. Dunn *et al.*, *Angew. Chem., Int. Ed. Engl.*, 1997, **36**, 645.

Topic 23
Nitric Oxide

R. M. J. Palmer *et al.*, *Nature*, 1987, **327**, 524.

A. R. Butler *et al.*, *Chem. Soc. Rev.*, 1993, **22**, 233.

R. Rossaint *et al.*, *Methods in Enzymology*, 1996, **269**, 442.

Topic 24
Taxol

M. C. Wani *et al.*, *J. Am. Chem. Soc.*, 1971, **93**, 2325.

K. C. Nicolaou *et al.*, *Angew. Chem., Int. Ed. Engl.*, 1994, **33**, 15.

R. A. Holton *et al.*, *J. Am. Chem. Soc.*, 1994, **116**, 1597.

Topic 25
Organometallic Complexes of C_{60}

P. Fagan *et al.*, *Science*, 1991, **252**, 1160.

A. L. Balch *et al.*, *Inorg. Chem.*, 1994, **33**, 5238.

H. F. Hsu *et al.*, *J. Am. Chem. Soc.*, 1996, **118**, 9192.

Topic 26
Symmetric Chlorine Dioxide

R. H. Derby *et al.*, *Inorg. Synth.*, 1953, **4**, 152.

S. Hubinger *et al.*, *Chem. Phys.*, 1994, **181**, 247.

A. L. Schmeltekopf *et al.*, *J. Geophys. Res.*, 1987, **92**, 8329.

Topic 27
Dendrimers

D. A. Tomalia *et al.*, *Angew. Chem., Int. Ed. Engl.*, 1990, **29**, 138.

C. J. Hawker *et al.*, *J. Am. Chem. Soc.*, 1990, **112**, 7638.

H. B. Mekelburger *et al.*, *Angew. Chem., Int. Ed. Engl.*, 1992, **31**, 1571.

Topic 28
Ascaridole

R. A. Bernhard *et al.*, *Phytochemistry*, 1971, **10**, 177.

G. O. Schenck *et al.*, *Annalen* 1953, **584**, 125.

R. Srininvasan *et al.*, *J. Am. Chem. Soc.*, 1979, **101**, 7424.

Topic 29
Artemisinin

A. R. Butler *et al.*, *Chem. Soc. Rev.*, 1992, **21**, 85.

Qinghausu Research Group, *Sci. Sinica*, 1980, **23**, 380.

M. A. Avery *et al.*, *J. Am. Chem. Soc.*, 1992, **114**, 974.

Topic 30
The Wacker Process

J. Smidt *et al.*, *Angew. Chem.*, 1959, **71**, 176.

J. Smidt *et al.*, *Angew. Chem.*, 1962, **1**, 80.

J. Tsuji *et al.*, *Synthesis*, 1984, **5**, 369. (BIDS refer to journal as *Synthesis-Stuttgart*)

Appendix C

The spectra for the Hwuche-Hwuche Bark exercise

Sample A: Mass spectrum

Elemental Analysis
C: 74.49%
H: 7.39%
N: 8.58%

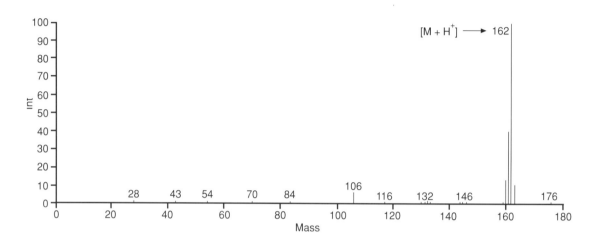

Sample A: Infrared spectrum

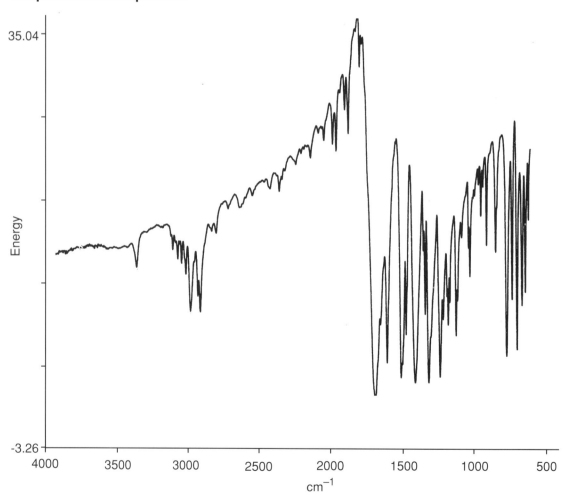

Sample A: ^{1}H NMR Spectrum (δ_H, 200 MHz, CDCl$_3$)

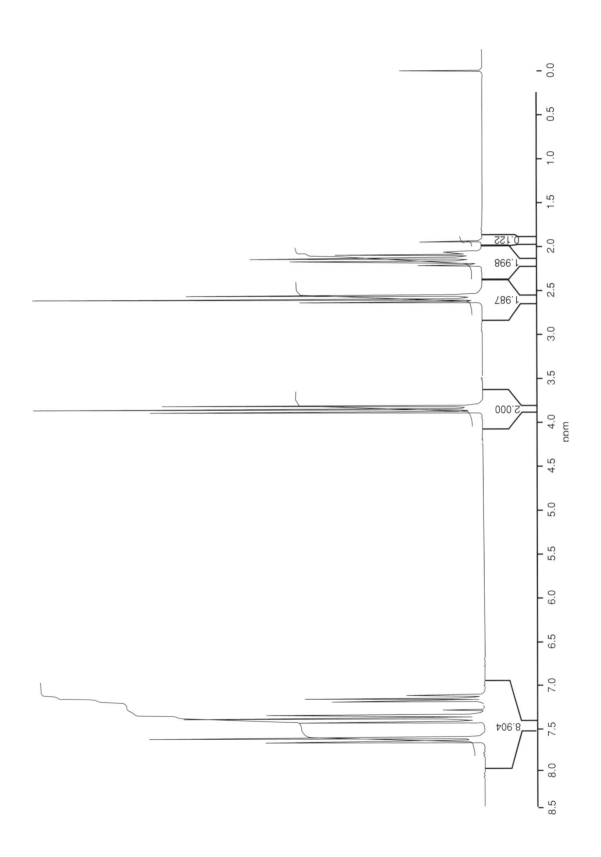

Sample A: ^{13}C NMR spectra (δ_c, 50 MHz, $CDCl_3$)

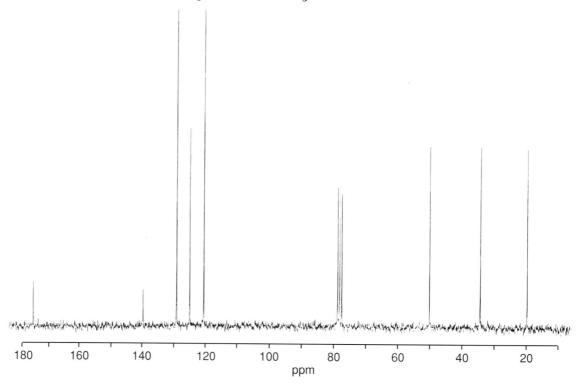

Sample B: Mass spectrum

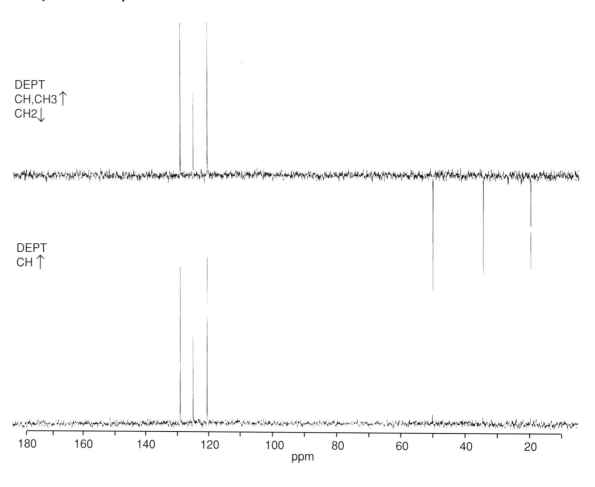

DEPT
CH,CH3 ↑
CH2 ↓

DEPT
CH ↑

Sample B: Infrared spectrum

High resolution mass spectrum for M⁺@161:161.0852

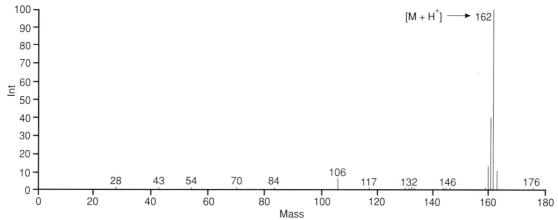

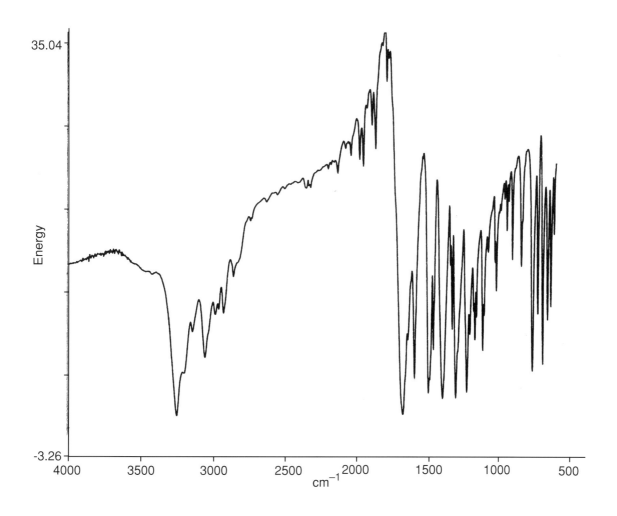

Sample B: ^{1}H NMR spectra (δ_H, 200 MHz, CDCl$_3$)

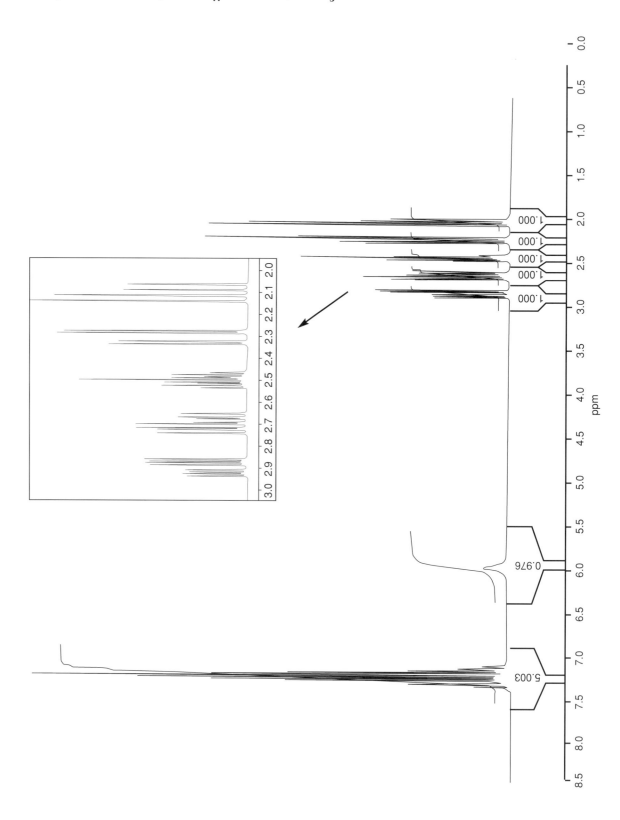

Sample B: ^{13}C NMR spectra (δ_c, 50MHz, CDCl$_3$)

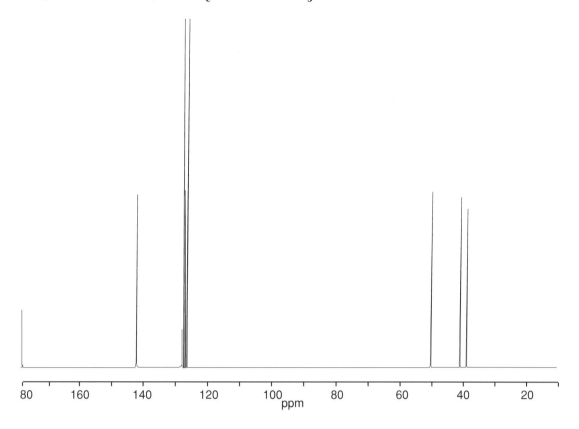

DEPT
CH,CH3 ↑
CH2 ↓

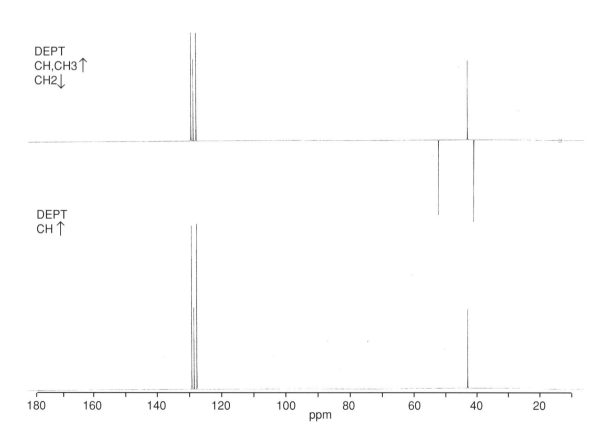

DEPT
CH ↑

Sample C: Mass spectrum

High resolution mass spectrum for M$^+$@165:165.0807

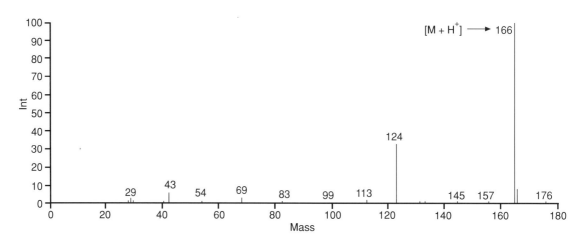

Sample C: Infrared spectrum

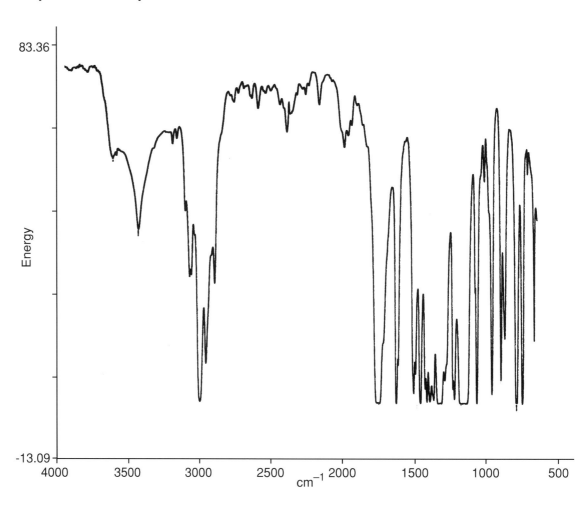

Sample C: ^{1}H NMR spectrum (δ_H, 200 MHz, CDCl$_3$)

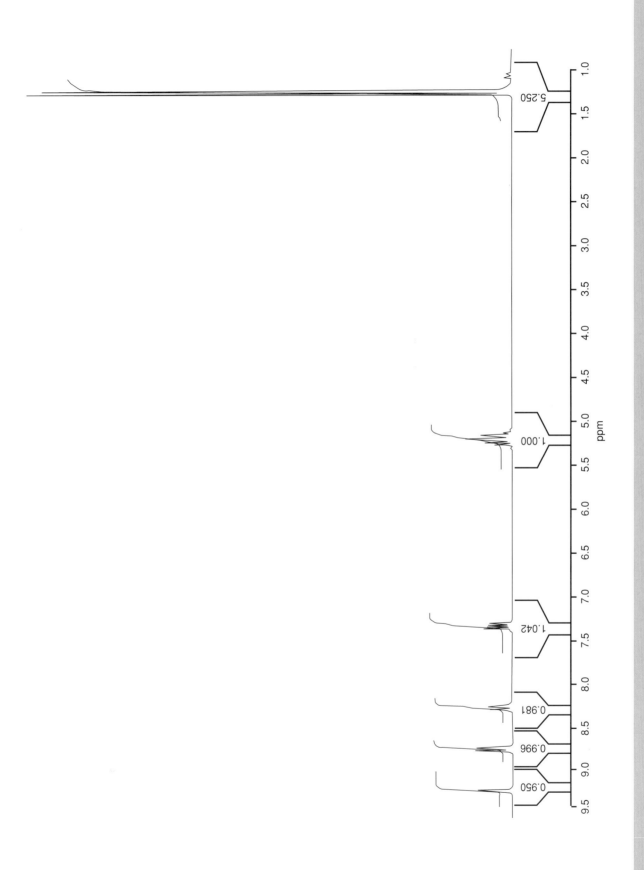

Sample C: ^{13}C NMR spectra (δ_c, 50MHz, CDCl$_3$)

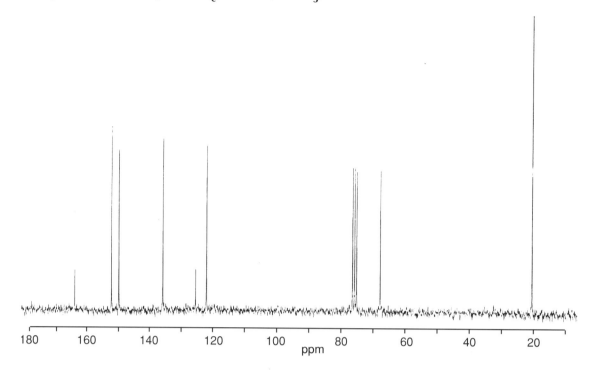

DEPT
CH,CH3 ↑
CH2 ↓

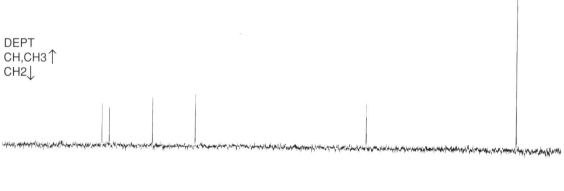

DEPT
CH ↑

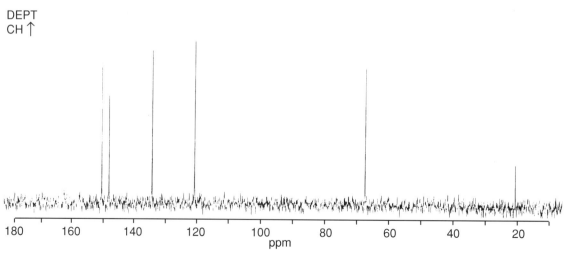

Sample D: Mass spectrum

Elemental analysis:
C: 75.84%
H: 7.92%
N: 7.95%

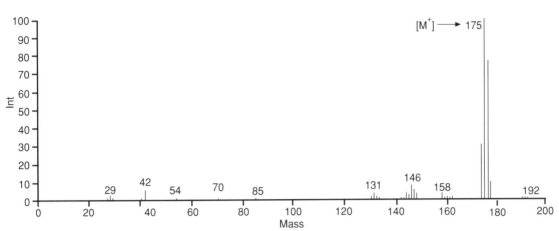

Sample D: Infrared spectrum

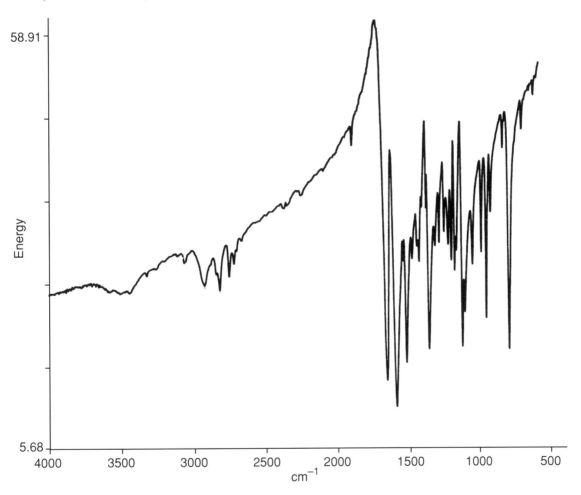

Sample D: ^{1}H NMR spectrum (δ_{H}, 200 MHz, CDCl$_3$)

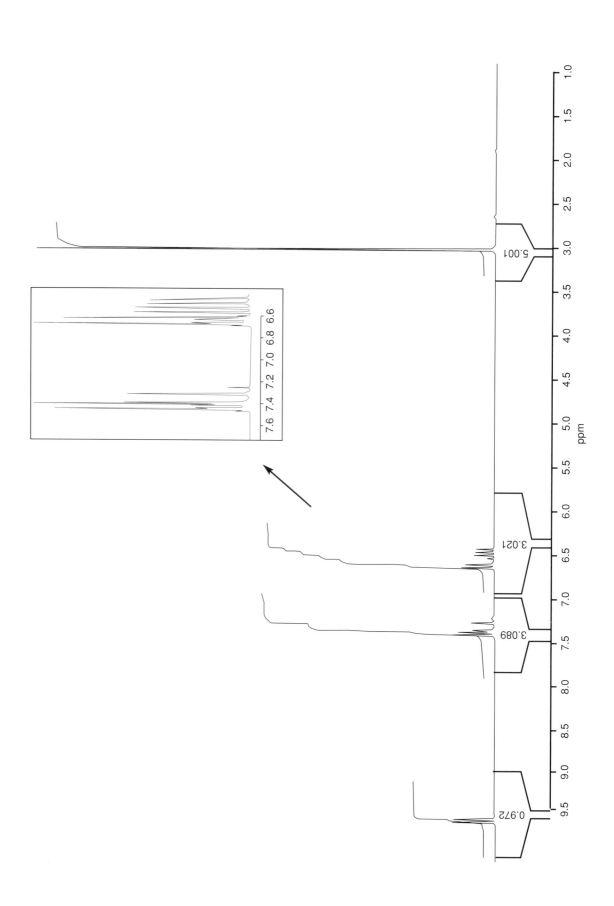

Sample D: ^{13}C NMR spectra (δ_c, 50MHz, CDCl$_3$)

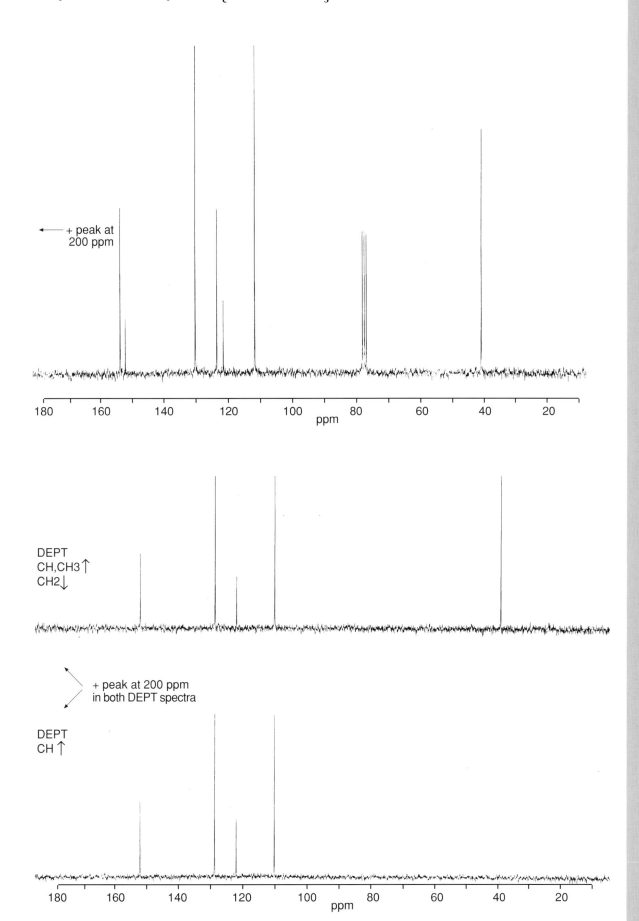

← + peak at
200 ppm

DEPT
CH,CH3 ↑
CH2 ↓

+ peak at 200 ppm
in both DEPT spectra

DEPT
CH ↑

Sample E: Mass spectrum

Elemental analysis:
C: 73.83%
H: 8.75%
N: 8.46%

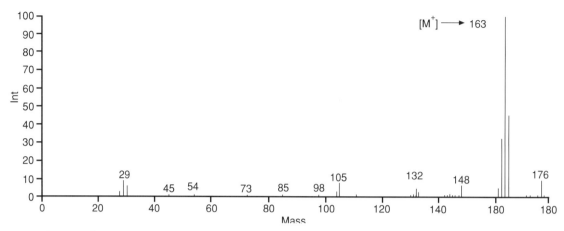

Sample E: Infrared spectrum

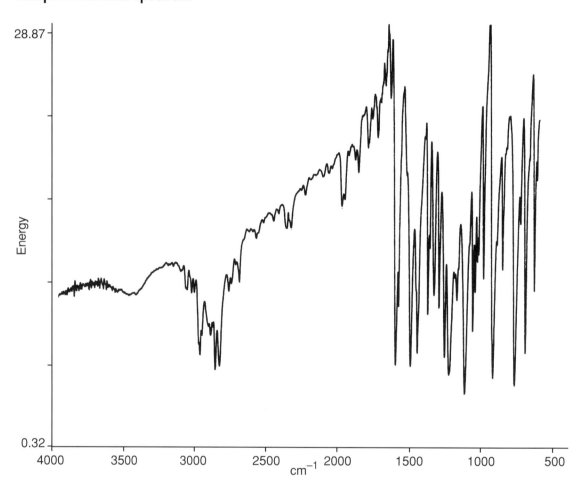

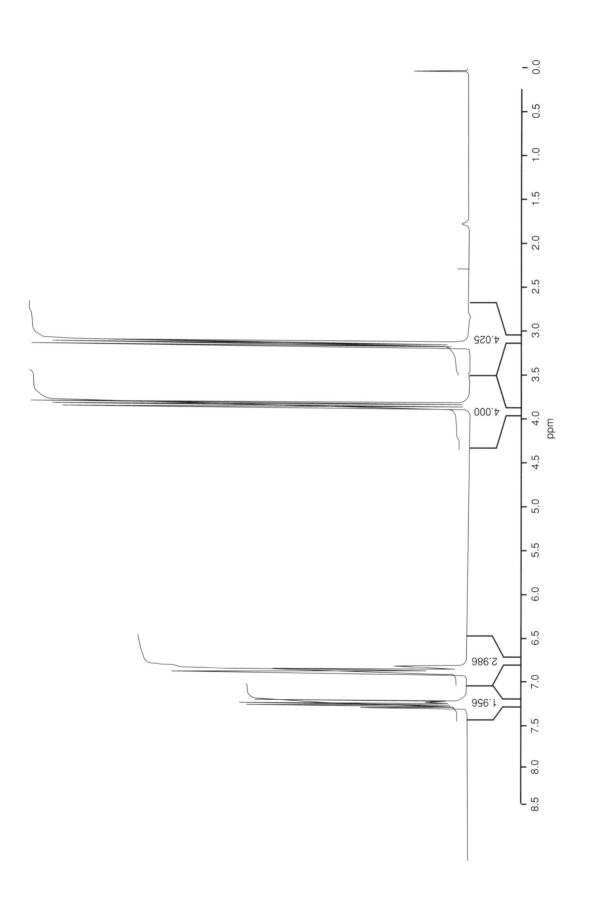

Sample E: ^{13}C NMR spectra (δ_c, 50MHz, CDCl$_3$)

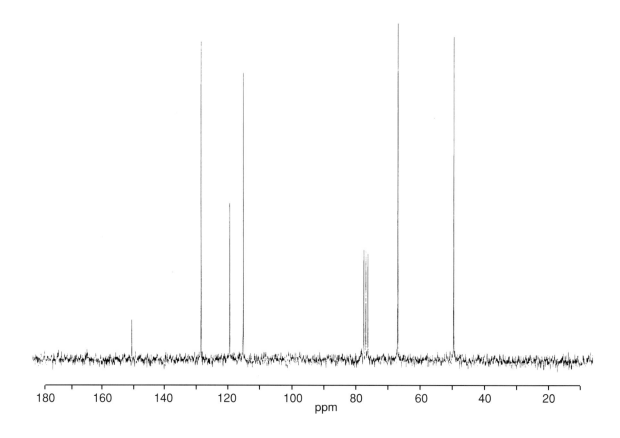

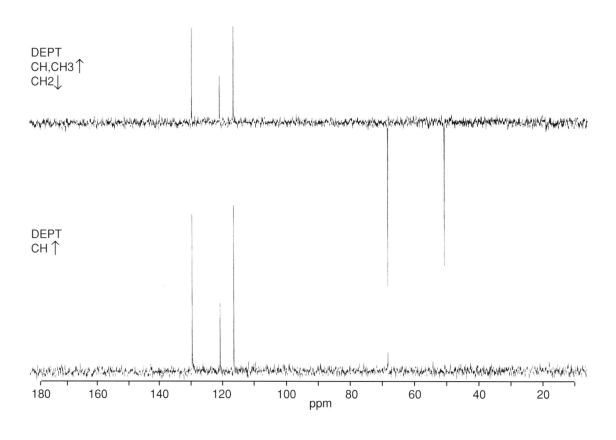

Sample F: Mass spectrum

Elemental analysis:
C: 73.92%
H: 8.51%
N: 8.53%

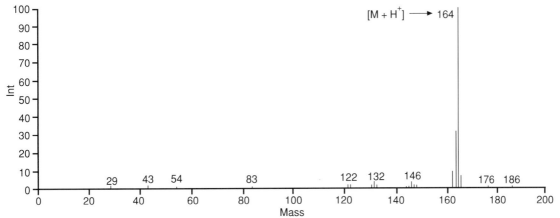

Sample F: Infrared spectrum

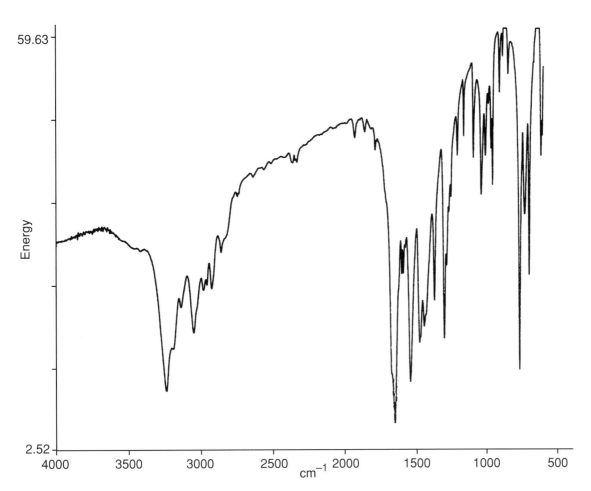

Sample F: ^{1}H NMR spectrum (δ_H, 200 MHz, CDCl$_3$)

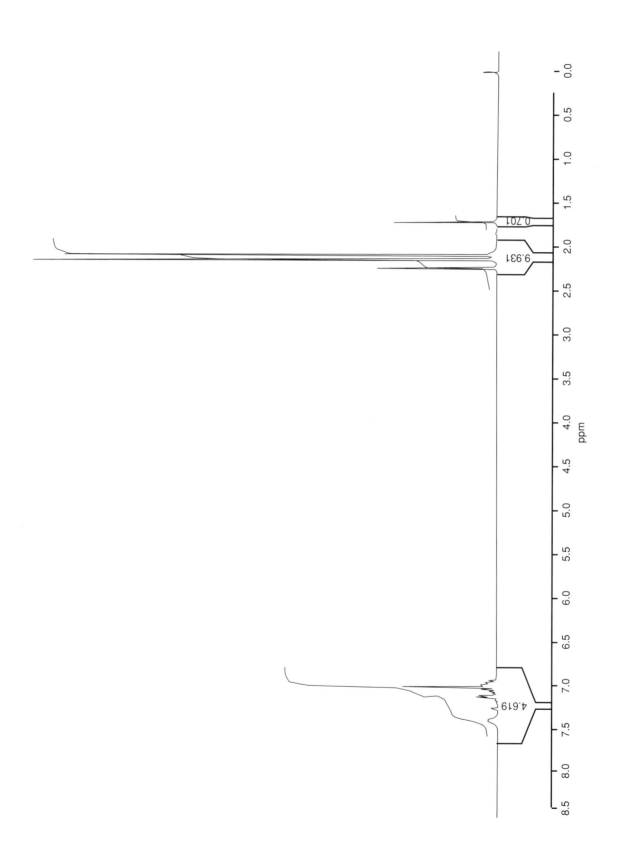

Sample F: ^{13}C NMR spectra (δ_c, 50MHz, CDCl$_3$)

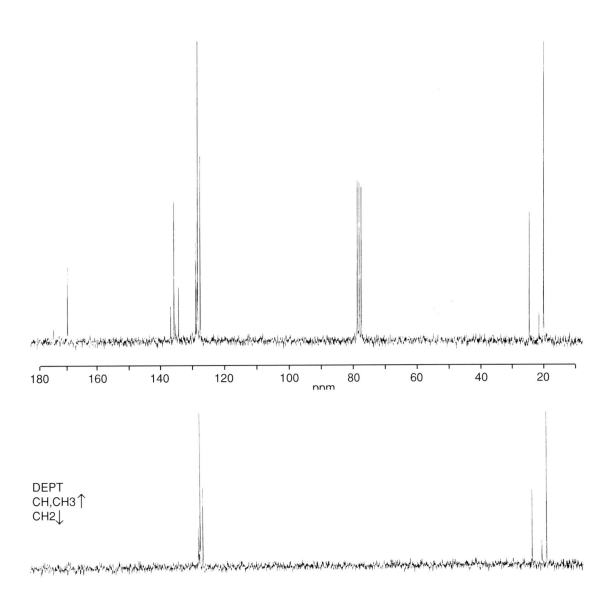

DEPT
CH,CH3 ↑
CH2↓

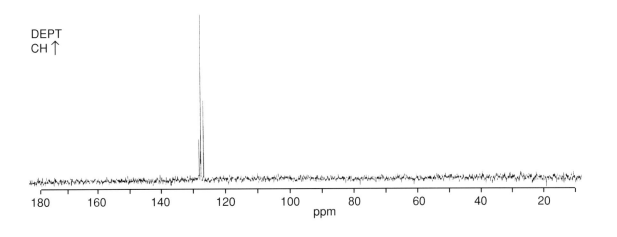

DEPT
CH ↑

Appendix D

Structure A

Elemental analysis and mass spectrum give molecular
formula $C_{10}H_{11}NO$.
Double bond equivalents = 6
4DBE in benzene ring, 1 lactam ring, 1 carbonyl

IR spectrum
Key features:
$3000\ cm^{-1}$	CH stretch
$1700\ cm^{-1}$	CO (carbonyl) stretch
$800–600\ cm^{-1}$	Aromatic CH deformation

Amides normally appear at $<1700\ cm^{-1}$, but the five membered lactam ring
is seen at $1700\ cm^{-1}$. The strong doublet feature in the fingerprint region is
characteristic of a mono-substituted aromatic ring.

^{1}H NMR spectrum
Key features:
δ 2.2, 2H, quin
δ 2.6, 2H, t
δ 3.8, 2H, t
δ 7.5, 5H, m

Aromatic protons arranged in a 2:2:1 ration confirming mono-substituted
ring. Coupling of aliphatic protons shows two triplets (coupled to CH_2) and
one complex triplet of triplets (coupled to two CH_2).

^{13}C NMR spectrum
Key features:
174 ppm	Amide carbon
140 ppm	Weak aromatic signal
130–120 ppm	Three signals, showing five aromatic carbons on integration
77 ppm	Solvent signal
44,33,18 ppm	Three aliphatic carbons

Weak signal at 140 ppm from quarternary carbon, and three additional
aromatic signals indicate a mono-substituted ring. The DEPT spectrum
confirms three CH_2 and five aromatic CH.

Mass spectrum
Key features:
161 m/z	Molecular ion
84 m/z	Loss of C_6H_5
77 m/z	Loss of lactam ring

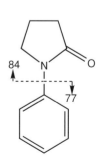

Structure B

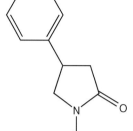

High resolution MS givees molecular formula
$C_{10}H_{11}NO$
Double bonds equivalents = 6
4DBE in benzene ring, 1 lactam ring, 1 carbonyl

IR spectrum

Key features:
3600 cm^{-1}	NH stretch
3000 cm^{-1}	CH stretch
1700 cm^{-1}	CO (carbonyl) stretch
800–600 cm^{-1}	Aromatic CH deformation

Amides normally appear at <1700 cm^{-1}, but the five membered lactam ring is seen at 1700 cm^{-1}. The strong doublet feature in the fingerprint region is characteristic of a mono-substituted aromatic ring.

^{1}H NMR spectrum

Key features:
δ 2.1, 1H, m
δ 2.3, 1H, m
δ 2.5, 1H, m
δ 2.7, 1H, m
δ 2.9, 1H, m
δ 6.0, 1H, br
δ 7.1–7.4, 5H, m

Coupling constants provide definitive information on the structure. Large coupling constants (J = 10 Hz) are most likely for geminal (H–C–H) protons. The patterns of intensities and the geminal coupling constant (J = 10 Hz) indicate that the signals at δ2.9 and 2.7 are one CH$_2$ group and that those at 2.3 and 2.1 are another CH$_2$ group. The CH at δ2.5 is complex (NOT a dd) so must be adjacent to 2 CH$_2$ groups, hence indicating that the Ph ring is attached to the 4-position.

^{13}C NMR spectrum

Key features:
178 ppm	Amide carbon
142 ppm	Aromatic signal
130–120 ppm	Three signals, showing five aromatic carbons on integration
50,41,39 ppm	Three aliphatic carbons

Aromatic carbons in four environments indicate mono-substituted ring. The DEPT spectrum confirms two CH$_2$, one CH and five aromatic CH.

Mass spectrum

Key features:
161 m/z	Molecular ion
84 m/z	Loss of C_6H_5
77 m/z	Loss of lactam ring

Structure C

High resolution MS gives the molecular formula $C_9H_{11}NO_2$
Double bond equivalents = 5
4DBE in benzene ring, 1 carbonyl

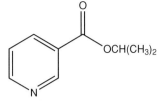

IR spectrum

Key features:

3000 cm^{-1}	CH stretch
1730 cm^{-1}	CO (carbonyl) stretch
1400 cm^{-1}	CN stretch
700 cm^{-1}	Aromatic CH deformation

Amides normally appear at $<1700 \text{ cm}^{-1}$, and there is no NH stretch. The IR spectrum therefore eliminates the presence of an amine, and confirms an ester.

^{1}H NMR

Key features:
δ 1.3, 6H, d
δ 5.3, 1H, septet
δ 7.4, 1H, dd
δ 8.3, 1H, br d
δ 8.8, 1H, br d
δ 9.2, 1H, s

The aromatic protons are shifted downfield, suggesting proximity to a nitrogen. The singlet at δ 9.2 gives the substitution pattern on the benzene ring. The six proton doublet at δ 1.3 strongly suggests an isopropyl group. The multiplet at δ 5.3 suggests a single CH appearing downfield due to its proximity to oxygen.

^{13}C NMR

Key features:

164 ppm	Ester carbon
153–123 ppm	Aromatic signals (one very weak at 127 ppm)
77 ppm	Solvent signal
69 ppm	Aliphatic carbon
22 ppm	Two equivalent aliphatic carbons

Aromatic carbons shifted downfield by nitrogen. Weak signal at 127 ppm from quarternary carbon (not seen in the DEPT spectrum). Isopropyl ester confirmed by two equivalent CH_3 and one CH.

Mass spectrum

Key features:

165 m/z	Molecular ion
150 m/z	Loss of CH_3
106 m/z	Loss of $OCH(CH_3)_2$
87	Isopropyl ester group

Structure D

Elemental analysis and MS give the
molecular formula $C_{11}H_{13}NO$
Double bonds equivalents = 6
4DBE in benzene ring, 1 double
bond, 1 carbonyl

IR spectrum
Key features:

3000 cm^{-1}	CH stretch
1700 cm^{-1}	CO (carbonyl) stretch
1640 cm^{-1}	Double bond (conjugated)
$800, 700 \text{ cm}^{-1}$	Aromatic CH deformation

The stretch at 1700 cm^{-1} is not an amide, which would appear at
$<1700 \text{ cm}^{-1}$, but an aldehyde or ketone. The aromatic features in the
fingerprint region are characteristic of a para-substituted benzene ring.

^{1}H NMR
Key features:

δ 3.0, 6H, s
δ 6.4–6.7, 3H, m
δ 7.3–7.5, 3H, m
δ 9.6, 1H, d

Peaks from the aromatic and aliphatic protons can be identified from the
coupling constants and splitting patterns. The aldehyde doublet (δ9.5) is
vicinally coupled to one alkene proton (J=8 Hz); the alkene proton is at δ6.5.
This proton is also coupled to the second alkene proton (J = 15.6 Hz). The
same coupling constant is found for the signal at δ7.3, identifying this as the
second alkene proton; the large coupling constant identifies the alkene or
trans. The aromatic protons are at δ7.5 and δ6.7.

^{13}C NMR
Key features:

200 ppm	Aldehyde carbon
155–110 ppm	Aromatic and alkene carbons
77 ppm	Solvent signal
40 ppm	Two equivalent aliphatic carbons

The DEPT spectrum identifies the two quarternary aromatic carbons at 153
ppm and 121 ppm. The two equivalent carbons at 40 ppm are missing from
the lower spectrum, so they are confirmed as methyl carbons. The two pairs
of equivalent CHs at 111 and 130 ppm indicate a para-substituted benzene
ring. The alkene carbons at 123 and 154 ppm are shifted downfield by their
proximimity to the aldehyde and aromatic system.

Mass spectrum
Key features:

175 m/z	Molecular ion
160 m/z	Loss of CH_3
131	Loss of $N(CH_3)_2$
120	$C_6H_5N(CH_3)_2$

Structure E

Elemental analysis and MS give molecular formula $C_{10}H_{13}NO$
Double bonds equivalents = 5
4DBE in benzene ring, 1 additional ring

IR spectrum

Key features:

~3000 cm^{-1}	CH stretches
700 cm^{-1}	Aromatic CH deformation

Note the lack of hydroxyl, amide or carbonyl.

^{1}H NMR spectrum

Key features:

δ 3.1, 4H, m
δ 3.9, 4H, m
δ 6.8, 3H, m
δ 7.3, 2H, m

The signals at δ3.1 and δ3.9 indicate two pairs of equivalent CH$_2$ groups, next to electron withdrawing N and O respectively; only the morpholine ring is consistent with this.

^{13}C NMR spectrum

Key features:

151 ppm	Aromatic carbon
123 ppm	Two equivalent aromatic carbons
120 ppm	One aromatic carbon
116 ppm	Two equivalent aromatic carbons
77 ppm	Solvent signal
66 ppm	Two equivalent aliphatic carbons
49 ppm	Two equivalent aliphatic carbons

Quarternary carbon at 151 ppm. All aliphatic carbons are CH$_2$.

Mass spectrum

Key features

163 m/z	Molecular ion
86 m/z	C_4H_8NO
77 m/z	C_6H_5

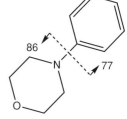

Structure F

Elemental analysis and MS give molecular
formula $C_{10}H_{13}NO$
Double bonds equivalents = 5
4DBE in benzene ring, 1 carbonyl

IR spectrum

Key features:
3600 cm^{-1} Broad NH stretch
1680 cm^{-1} CO (carbonyl) stretch
800 cm^{-1} Aromatic CH deformation

The value of the carbonyl stretch suggests the presence of an amide.
Fingerprint region suggests three adjacent aromatic protons.

^{1}H NMR spectrum

Key features:
δ 2.1, 3H, s
δ 2.2, 6H, s
δ 6.9–7.2, 3H, m
δ 7.4, 1H, br s
(δ1.7 and δ2.3; impurities)

All aliphatic protons appear very close due to electron withdrawing
environments (carbonyl and benzene ring).

^{13}C NMR spectrum

Key features:

169 ppm	Amide carbon
136 ppm	Aromatic carbon
128–130 ppm	Three aromatic carbons
77 ppm	Solvent signal
23 ppm	One aliphatic carbon
19 ppm	Two equivalent aliphatic carbons

Quarternary carbon at 169 ppm. All aliphatic carbons are CH_3.

Mass spectrum

Key features:

163 m/z	Molecular ion
148 m/z	Loss of CH_3
105 m/z	Loss of $NHCOCH_3$
58 m/z	$NHCOCH_3$
43 m/z	NCHO

Appendix E – Assessment of module

All of the exercises in this teaching pack can be run in isolation, but there are many advantages with running several of them as a "module" as outlined in the introduction. If this is the case, it is useful to have an overall assessment and feedback mechanism, as well as considering separate feedback for each of the exercises (see tutor's notes). The three forms in this Appendix outline one mechanism for gaining feedback, generating an overall assessed grade, and encouraging self-appraisal by students.

Information for students – feedback

The emphasis of this module is on practising communication skills, and receiving constructive feedback. Some of the feedback will be in the form of comments from tutors and colleagues, and some will be in the form of specific marks. There will be an overall mark for this module but, much more importantly, we hope you will benefit from the comments you will receive. At the end of the module, you will need to complete and return this form on team-work; then, after you have received all of the feedaback and marks, we want you to complete a self-assessment form for your own future reference.

Team work

Rate the contribution of the other mebers of your company working on The Fluorofen Problem, Scientific Paper, Hwuche-Hwuche Bark, Interviews and Poster Presentation. Your comments will be anonymous.

Your name _____

Your company name _____

How much did your colleagues contribute to the team work?

	Name	Loads	Their fair share	Not much	Nothing
1.					
2.					
3.					
4.					

Who was your partner for the *Chem Comm* article and DInCh entry?

How did you feel your contributions were divided up? (tick one)

I did it all
I did most of it
It was about 50:50
S/he did most of it
S/he did all of it

Overall, how well did you contribute to team work? Which parts of it were you good at, and which were you less good at?

Feeback for Communicating Chemistry

You will get the following feedback for the module, plus some additional comments on specific exercises:

1.	Computer keyboard skills	Grade
2.	WWW Treasure hunt	Mark
3.	*New Chemist* article	Mark
4.	'DInCh' entry	Grade
5.	Hwuche-Hwuche bark	Team mark
6.	Annual review presentation	Feedback
7.	Interviews	Feedback at the time
8.	Poster	Team mark

Overall effort (attendance and private study)
Participation and team work
General comments

Overall module grade

Marks and grades roughly equate to: A* = excellent = 85%; A = very good = 75%; B = good = 65%; C = 55% = OK; D = 45% = poor; E = 40% = barely adequate; X = 0% = awful. You need to pass everything.

Final feedback

Most of you have taken part enthusiastically in 'Communicating Chemistry" and we hope you have gained much from it. The module is not one where we aim primarily to teach you or assess you – the main aim is to give you the opportunity to try out some aspects of communication skills, and to think more about assessing yourself... peer review should have helped you to get a feel for doing this. You should now have a good idea of what you are good at, what you should avoid, and how you might improve your skills at communicating chemistry.

Lots of you really showed unexpected talents of initiative, team leadership or presentation skills. Nevertheless, one or two people in every exercise muddled something up – forgot to put their name on a submission, went to the wrong room, were late, or produced a scrappy CV... however great the excuse, it's worth remembering that you'll impress people far more if you get things right first time, and on time! Overall, we were delighted by how well you did, and we hope you managed to surprise yourselves too with some "hidden" talents.

Enclosed in this envelope is your final feedback. The letters mean the following: A = excellent; B = very good; C = quite good; D = poor; E = barely adequate; X = unacceptable. Whilst we would sooner not give an overall mark, it does seem fair to give proper credit for hard work and success – everyone who had taken part fully and completed the assignments will pass the module. However, we will enter an actual module mark using the following conversion:

A* = excellent = 85%
A = very good = 75%
B = good = 65%
C = OK = 55%
D = poor = 45%
E = barely adequate = 40%
X = not good enough or absent = 0%

To help you develop your communication skills, you should complete the "Final Self-Assessment" form enclosed. Don't hand it in as it is intended as your own summary of how you did in the module, to help you improve your communication skills. We hope the module has been useful and enjoyable, and good luck in using and developing your communication skills.

1.	The Fluorofen Problem	See teamwork*
2.	Scientific Paper	See teamwork*
3.	Computer Keyboard Skills	Grade:
4.	WWW treasure hunt	Success: /7
5.	*New Chemist* article	Mark:
		Comments:
6.	*DInCh*	Mark:
		Comments
7.	Hwuche-Hwuche Bark*	Mark:
8.	Annual Review Presentation	Feedback already provided
9.	Interviews*	Feedback provided at the time
10.	Posters*	Mark:

Effort (attendance and private study):
Participation and team work*:
General comments:

Overall module grade:

Final self assessment

Your name: _____

Summarise what you gained from each of the exercises – these are skills that you may wish to a) use again; b) tell prospective employers you have; c) improve upon.

1. The Fluorofen Problem

_____

2. Scientific Paper

_____

3. Computer keyboard skills [mark: _____]

_____

4. WWW Treasure Hunt [mark: _____]

_____

5. *New Chemist* article

_____

6. Dictionary of Interesting Chemistry [mark: _____]

_____

7. Hwuche-Hwuche Bark [mark: _____]

_____

8. Annual Review Presentation
Feedback comments _____

9. Poster presentation [mark: _____]

_____

10. Interviews
Feedback comments _____

State up to three things you felt you did particularly well

i) _____

ii) _____

iii) _____

State up to three things you feel you could improve upon

i) _____

ii) _____

iii) _____

Rate your ability:	Great	Good	OK	Weak	Awful
Writing clear consie reports?					
Computer keyboard skills?					
Computer database skills?					
Explaining things to others in writing?					
Problem solving?					
Working in a team?					

What skills do you especially plan to develop in the near future?